2ND EDITION

Lake Superior Rocks & Minerals

Field Guide

Dan R. Lynch & Bob Lynch

Adventure Publications
Cambridge, Minnesota

Dedication

Dan dedicates this work to his wife, Julie, for her unending support for this and every other book he's written, and to his parents, Bob and Nancy, for all that they do for him. Also, thanks to Wes Lynch, Dan's brother, for his blunt and honest critique of the first edition, which bettered the way Dan has approached all the books he's written since.

Acknowledgments

Thanks to Jim Cordes, Chris Cordes, John Woerheide, Dave Woerheide, George Robinson (formerly of the A.E. Seaman Mineral Museum), Mark Bowan, Phil Burgess, Eric Powers, Alex Fagotti, Bradley A. Hansen, David Gredzens, Daniel Bubalo Jr., and especially Bob Wright for specimens and information.

All photography by Dan R. Lynch except page 17 by Emily Lundstrom/ Shutterstock.

Cover and book design by Jonathan Norberg

Edited by Brett Ortler

10 9 8 7 6 5 4 3

Lake Superior Rocks & Minerals
First Edition 2008, Second Edition 2022

Published by Adventure Publications
An imprint of AdventureKEEN
310 Garfield Street South
Cambridge, Minnesota 55008
(800) 678-7006
www.adventurepublications.net

Printed in China
ISBN 978-1-64755-058-5 (pbk.); 978-1-64755-059-2 (ebook)

TABLE OF CONTENTS

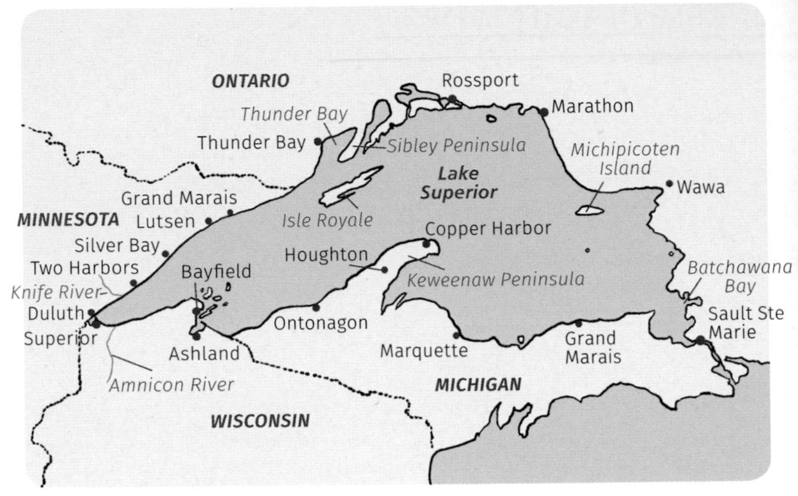

INTRODUCTION

Anyone who has spent time on Lake Superior knows that it is more than a mere lake. As the world's largest lake by surface area, Lake Superior is more akin to a sea, and the wave-battered rocks along its shores can attest to its incredible scale and power. But those rocks are much, much older than the mighty lake—most are a billion years old or more—and it is within them that many of the region's most famed collectibles formed. From agates to copper and silver, Lake Superior's rugged shores and rivers can yield once-in-a-lifetime specimens that would thrill anyone, and thanks to the glaciers of the past Ice Ages, such finds may be just sitting on the surface, waiting to be found. Lake Superior is a destination unlike any other, thanks to the region's dramatic geological history, its picturesque backdrop of soaring cliffs and wild waves, and the potential for stunning finds.

A NOTE ABOUT THE 2ND EDITION

This edition of *Lake Superior Rocks & Minerals* incorporates numerous changes and improvements over the 2008 original. The text is entirely rewritten and is complemented by hundreds of new photos that cover a wider range of rocks and minerals. We have also redefined the region covered by the book. This edition is more closely focused on Lake Superior's shores and omits mineral species that are only found further inland; in general, this book will be most relevant to areas within 10 to 20 miles of Lake Superior. Beyond that range, this book will still remain useful, but you may begin to encounter other rocks not covered here. This way, we can better discuss all the material you'll find walking the shores or rivers of Lake Superior itself. Like the first edition, this book covers Minnesota, Wisconsin, and Michigan, but it also now includes Lake Superior's shores in Ontario, Canada, making this the complete around-the-lake rock and mineral field guide. Lastly, we will not be covering Isle Royale or Michipicoten Island here because both islands are federally protected and all collecting is illegal; we encourage our readers to help preserve these remote and wild places by respecting these laws.

We welcome new readers to explore this amazing region with us, and we thank readers of the first edition for your continued support and hope to expand your knowledge with this new edition.

IMPORTANT TERMS AND DEFINITIONS

Geology and mineralogy are topics full of terms that may initially seem complicated but are important to your understanding of the sciences. So to make this book educational for novices yet still useful for experienced collectors, we have included some technical terms in the text but we "translate" them immediately after by providing a brief definition. In this way, amateurs can learn some of the more important terms relevant to the hobby in an easy, straightforward manner. Of course, all of the geology-related terms used here are defined in the glossary found at the back of this book as well. But for those entirely new to rock and mineral collecting, there are a few very important terms you should understand not only before you begin researching and collecting minerals, but even before you read this book.

Many people may begin hunting for rocks and minerals without knowing the difference between the two. A **mineral** is the crystallized (solidified) form of an inorganic chemical compound, or combination of elements. For example, silicon dioxide, a chemical compound consisting of the elements silicon and oxygen, crystallizes, or hardens, to form quartz, one of the most abundant minerals on Earth. In contrast, a **rock** is a mass of solid material containing a mixture of many different minerals. While pure minerals exhibit very definite and testable characteristics, such as a distinct repeating shape and hardness, rocks do not and can vary greatly because of the various minerals contained within them. This can often make identification of rocks more difficult for amateurs.

Many of the important terms critical to rock hounds apply only to minerals and their crystals. A **crystal** is a solid object with a distinct shape and repeating atomic structure created when a chemical compound solidifies. In other words, when different elements come together, they form a chemical compound which will take on a very particular shape when it hardens. For example, the mineral pyrite is iron sulfide, a chemical compound consisting of iron and sulfur, which **crystallizes**, or solidifies, into the shape of a cube. A "repeating atomic structure" means that when a crystal grows, it builds upon itself. If you compared two crystals of pyrite, one an inch long and the other a foot long, they would have the same identical cubic shape. In contrast, if a mineral is not found in a well-crystallized form but rather as a solid, rough chunk comprised of tiny mineral grains, it is said to be **massive**. If a mineral typically forms **massively**, it will frequently be found as irregular pieces or masses, rather than as well-formed crystals.

Cleavage is the property of some minerals to break in a particular way when carefully struck. As solid as minerals may seem, many have planes of weakness within them derived from a mineral's molecular structure. These points of weakness are called **cleavage planes** and it is along these planes that some minerals will **cleave**, or separate, when struck. For example, the mineral galena has cubic cleavage, and even the most irregular piece of galena will fragment into perfect cubes if carefully broken.

Cleavage is a different geological property than fracture. **Fracture** is the shape or texture of a random crack or break in a rock or mineral. One of the most prominent examples discussed in this book is **conchoidal fracture**, which is the trait of some minerals to develop semi-circular or smooth, curving cracks when struck or broken. Some minerals can have both distinct fracture and cleavage patterns.

Luster is the intensity with which a mineral reflects light. The luster of a mineral is described by comparing its reflectivity to that of a known material. A mineral with "glassy" luster (also called "vitreous" luster), for example, is similar to the "shininess" of glass. The distinction of a "dull" luster is reserved for the most poorly reflective minerals, while "adamantine" describes the most brilliant (though is typically reserved for diamond). Minerals with a "metallic" luster clearly resemble metal, which can be a very diagnostic trait. But determining a mineral's luster is a subjective experience, so not all observers will necessarily agree, especially when it comes to less obvious lusters, such as "waxy," "greasy," and "earthy."

When minerals form, they do so on or in rocks. Therefore, it is important to understand the distinction between the different types of rocks if you hope to successfully find a specific mineral. **Igneous** rocks form as a result of volcanic activity and originate from magma, lava, or volcanic ash. **Magma** is hot, molten rock buried deep within the earth, and it can take extremely long periods of time to cool and harden to form a rock. **Lava**, on the other hand, is molten rock that has reached the earth's surface where it cools and solidifies into rock very rapidly. **Sedimentary** rocks typically form at the bottoms of lakes and oceans when sediment compacts and solidifies into layered masses. This sediment can contain organic matter as well as weathered fragments or grains from broken-down igneous rocks, metamorphic rocks, or other sedimentary rocks. Finally, **metamorphic** rocks develop when igneous, sedimentary, or even other metamorphic rocks are subjected to heat and/or pressure within the earth and are changed in appearance, structure, and mineral composition.

A BRIEF OVERVIEW OF THE GEOLOGY OF LAKE SUPERIOR

If you've visited any part of Lake Superior's rugged shores, you know that the region's amazing geology is on display everywhere you look; the lava cliffs of Minnesota's shores, Wisconsin's sandy Apostle Islands, Michigan's copper-rich Keweenaw Peninsula, and the incredibly old rocks of Ontario's expansive shoreline are just a mere sampling. Billions of years and numerous major geological events contributed to the region's unique beauty, but the Lake's very existence is owed to one event more than any other: the Midcontinent Rift.

The Midcontinent Rift

Around 1.1 billion years ago, all of Earth's continents were arranged together in one giant landmass we call Rodinia, and the region that would later become North America was then situated over the equator. As the tectonic plates—the massive sheets of rock that make up the earth's crust—beneath Rodinia began to move and separate, the continent started to split apart and a massive tear, called a rift, slowly opened across the landscape. The rock weakened, allowing volcanic activity to spring up along the length of the rift, and molten rock from deep within the earth rose to fill the void as it widened, cooling to form volcanic rock as it reached the surface. But ultimately, the rift failed. It's uncertain exactly why the spreading ceased, but a leading theory is that the Midcontinent Rift was unable to spread further apart due to stronger tectonic movements occurring farther to the east. The opposing movement pushed against the Midcontinent Rift and effectively "pinched" it, stopping it from spreading further. If the Midcontinent Rift had continued, the continent would have split into two parts separated by a sea, as is happening today with the Red Sea, which lies in an active rift system separating Saudi Arabia from Egypt.

As the volcanic rocks that were formed during the rifting event began to cool and the volcanic activity beneath them subsided, the rocks contracted and collapsed, slumping downward and creating a wide trough along the rift's length. This sunken body of rock then collected water, and with water comes sediment—the granular fragments of worn-down rocks and minerals. Over the next billion years, the sediments collecting in the basin hardened to form sedimentary rocks that continued to build up, burying all evidence of the Midcontinent Rift for eons.

The Midcontinent Rift (red) across the region today

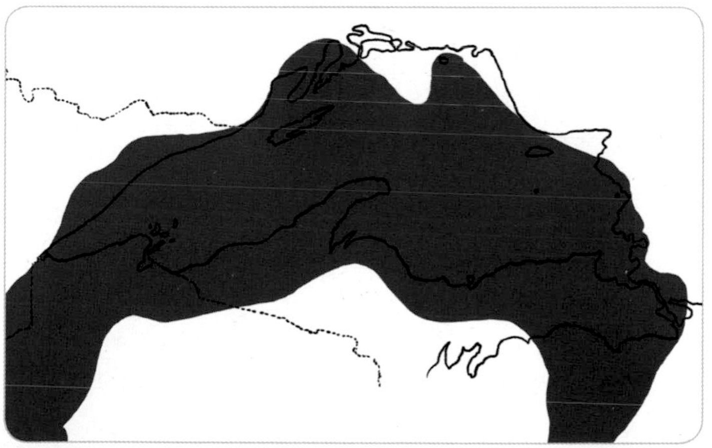

Hydrothermal Activity and Collectible Minerals

In most volcanically active regions, groundwater is heated and enriched with dissolved minerals and can then rise through the overlying rock as steam. This is called hydrothermal activity, and as the water percolates through existing rocks it can collect in cavities and cracks. As the water accumulates, the minerals within it can also accumulate and then crystallize. This process occurred during the Midcontinent Rift, and this is particularly relevant to collectors because many of Lake Superior's most desirable minerals formed this way, including agates, copper, and copper ores (ores are minerals that can be mined and processed to free the valuable metals they contain). Most of the time, the rock that hosts these collectible minerals is the same igneous rock produced in the volcanic event, but not always; in Michigan, copper is famously found deposited in older sedimentary rocks. And the heat and steam from hydrothermal activity not only deposits new minerals in rock cavities, but it can also begin to alter and change the rocks themselves.

The Ice Ages and the Glaciers

Over a billion years have passed since the Midcontinent Rift. In that time, the continents changed shape and shifted positions on the planet, life appeared on dry land, and later, the dinosaurs appeared and disappeared. Eventually, around 200,000 years ago, early humans began their great migration from Africa to all corners of the earth. All the while, the planet's climate was in constant flux. Much of this had to do with changes in the composition of Earth's atmosphere, causing both warm, wet periods as well as cold, icy ones. Periods of lower global temperatures are often called Ice Ages, and are characterized by glaciation, or the presence of glaciers.

Glaciers are enormous, thick sheets of ice that form when snow compacts, often near the poles or at high elevations, and then slowly grow in size. Every continent has seen glaciation, especially during the last Ice Age, and North America's ice sheets achieved incredible size. Their continued growth in the cold, northern reaches of Canada

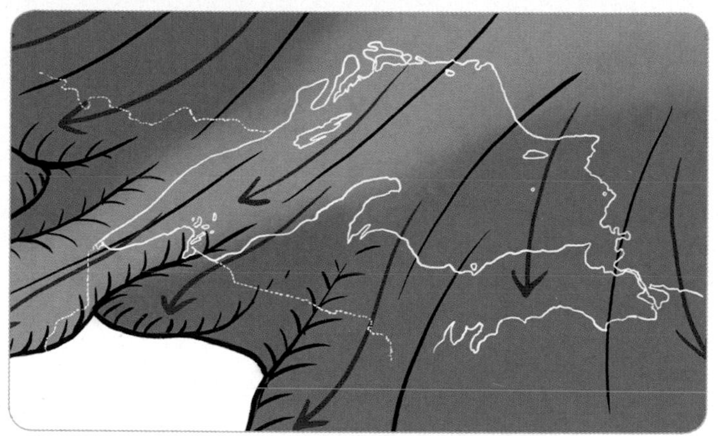

General path of glacial movement during the last Ice Age

caused them to “flow” slowly southward; as the climate warmed, they began to melt and retreat northward. There have been numerous Ice Ages on Earth, but we have the most evidence of the last one, which began around 110,000 years ago and lasted until just 10,000 years ago. The glaciers produced during this most recent Ice Age were up to a mile thick, and their immense weight crushed and pulverized the rock below them. As the crushed rock was incorporated into the ice, its abrasiveness increased, and over 100,000 years the repeated advance and retreat of the glaciers scoured away most of the soft sedimentary rocks that once filled the Midcontinent Rift basin. Then, as the Ice Age waned and global temperatures rose, the glaciers began to melt and retreat for the last time, leaving billions of gallons of water behind, which both formed the region’s countless small lakes and filled the great basin we know today as Lake Superior.

The glaciers are responsible for exposing many of the surface rocks in the region and, in turn, for making many of the Lake’s most collectible minerals accessible. As they scoured their way across the landscape, the glaciers removed all plant life and soft overlying rocks, revealing the hard, volcanic Midcontinent Rift rocks below. This is important to collectors because the Rift's rocks are the geological home to valuable collectibles like agates and copper. Without the excavation work done by the glaciers, these beloved collectibles would still be out of reach.

As the glaciers crushed underlying rocks, rock fragments were incorporated into the ice, causing the glaciers to spread rocks far and wide. This is why we find many Lake Superior rocks and minerals—especially agates—far from their original source; they were deposited when the glaciers melted. For example, many of the speckled cobbles you'll find all along Lake Superior's southern shores originated much farther north, in a massive rock formation known as the Canadian Shield.

The Canadian Shield

The Canadian Shield is an enormous area of exposed rock connecting the Great Lakes region to the Arctic Ocean. Representing the ancient core of North America and once the deeply buried "roots"

Canadian Shield (red)

of a long-gone mountain range, the nearly 5-million-square-mile expanse of rock was pushed up to the surface, and then deeply weathered and smoothed by glaciers. The rocks of the Canadian Shield are largely coarse-grained igneous rocks, such as granite, and metamorphic rocks derived from them. As a whole, the Shield is nearly 4 billion years old. It is through these ancient rocks that the Midcontinent Rift tried to separate, and much of the northern shores of Lake Superior are comprised of exposed Shield rocks today. Fragments of Canadian Shield rocks were also scattered far to the south by the glaciers.

Iron Ranges and Sedimentary Rocks

Long before the Midcontinent Rift occurred, an ancient sea covered what is now the Lake Superior region. Beginning around 2.5 billion years ago, sediments from Earth's barren, lifeless land readily washed into this sea and collected near its shores. Over time, the sediments—which were rich with iron—compressed and hardened, forming a hard, tightly layered sedimentary rock known as Banded Iron Formation (read more on page 93). These ancient, iron-rich shorelines are still present in the Lake Superior region today as Iron Ranges, most of which have seen some amount of mining activity in recent decades.

There are several Iron Ranges in the wider Lake Superior region, but only a few are close enough to the Lake to be relevant to area rock hounds: the Gunflint Range in northern Minnesota and southern Ontario, the Gogebic Range in Wisconsin and Michigan, the Marquette Range in Michigan, and the Michipicoten Range in Ontario (though the Michipicoten Range did not form the same way as the other ranges mentioned here, it still produces similar rocks and minerals). These areas are geologically important, both scientifically and economically, and will receive much attention in any book about Lake Superior's rocks.

There are some other important Lake Superior sedimentary rock formations worthy of note. The Copper Harbor Conglomerate, found

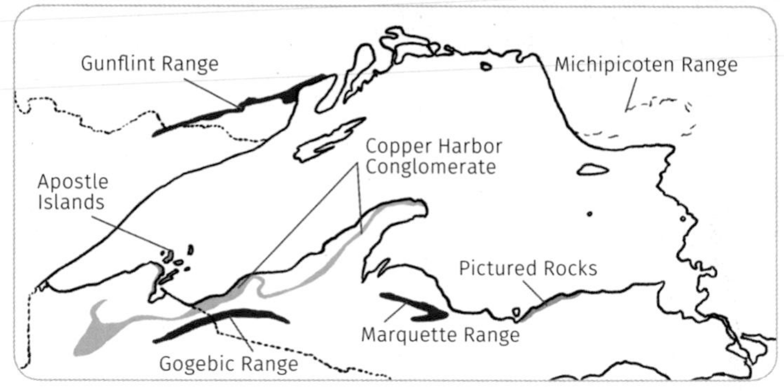

along the northwestern shores of the Keweenaw Peninsula in Michigan and extending into Wisconsin, formed from the rivers that once emptied into the basin formed by the Midcontinent Rift rocks. This rock formation consists of round cobbles and pebbles locked together by finer-grained sediments (a rock called conglomerate, see page 119 for more), and is important not only because of its size and prominence in Michigan, but also because later hydrothermal activity injected copper and other minerals into the rock. Similarly, further east in Michigan and Wisconsin, a younger body of sedimentary rock, known as the Nonesuch Shale, also saw the introduction of copper, which was deposited as broad, flat sheets between the layers in the rock.

Finally, while the glaciers scoured, crushed, and swept away untold amounts of soft sedimentary rocks, many were able to survive the Ice Ages. Two prominent examples are the Apostle Islands of Wisconsin and the Pictured Rocks National Lakeshore of Michigan. While both areas are nationally protected lakeshores where collecting is prohibited (with the exception of Madeline Island), they consist of incredible sandstone formations that were worn and sculpted by the glaciers and by Lake Superior itself, but they still survive to this day. These rock formations, which originated on ancient seafloors, are a stark contrast to the volcanically formed cliffs and ledgerock so common around Lake Superior.

While the geological processes and rock formations mentioned here are just a sampling of the incredible history of the region, they are all part of what makes Lake Superior a rock hound's destination. And thanks to the glaciers, much of the region's most desirable rocks and minerals are just lying on the surface, waiting to be found on your next walk along the water's edge.

PRECAUTIONS AND PREPARATIONS

It should come as no surprise that rock and mineral collecting can bring with it some dangers and legal concerns. It is always your responsibility to know where you can legally collect, which minerals may be hazardous to your health, and what to bring with you to ensure your safety. Here we will detail some of these issues.

Protected and Private Land

The Lake Superior region has several nationally protected parks and monuments as well as Native American reservations, and it is illegal to collect anything on those sites. For example, Isle Royale is a National Park, and Michipicoten Island is a Provincial Park, and both are federally protected, making collecting there illegal in all cases. Some state parks may allow collecting but only if you have obtained proper permits—others forbid collecting entirely. It is up to you to do the research before you go out hunting. We encourage collectors to obey the law and leave designated natural spaces wild and untouched for generations to come. It is always your responsibility to know whether or not the area in which you are collecting is protected.

As in any state, many places around Lake Superior are privately owned, including areas of wilderness that may not have obvious signage. Needless to say, you are trespassing if you collect on private property and the penalty may be worse than just a fine. In addition, property lines and owners change frequently, so just because a landowner gave you permission to collect on their property last year doesn't mean the new owner will like you on their property this year. Always be aware of where you are.

Dangers of Rivers, Rock Piles, and the Mighty Lake

When in the field, vigilance and caution are key to remaining safe. Many gravel pits and mine dumps present amazing collecting opportunities if you're granted permission to collect there, but may have large rock piles or pit walls that are unstable and prone to collapse. Never go beneath overhanging rock, and keep clear of unstable rock walls. Rivers also present their own dangers, and even though the water's surface may look calm, strong currents may be present. It doesn't take very much moving water to make you lose your footing completely. The same goes for Lake Superior itself; sudden rip-currents can sweep you off your feet and carry you hundreds of feet out into the lake. If you're wading in the lake, never go out to where the water is past your knees, and always step slowly and carefully as sudden deep spots are very common. Lastly, remember that Lake Superior is no mere pond; it is famously temperamental, and sudden wind, waves, and storms can arise seemingly out of nowhere. Do not under any circumstances canoe or kayak out into the lake if you do not have sufficient life-saving experience, especially sea-kayak experience. Every year, inexperienced tourists need to be rescued from sudden extreme weather or exposure to Lake Superior's frigid water.

Equipment and Supplies

When you set out to collect rocks and minerals, there are a few items you don't want to forget. No matter where you are collecting, leather gloves are a good idea, as are knee pads if you plan to spend a lot of time on the ground. If you think you'll be breaking rock, bring your rock hammer (not a nail hammer) and eye protection. If the weather is hot and sunny, take the proper precautions and use sunblock,

and bring sunglasses, a hat, as well as ample drinking water. Lastly, bringing a global positioning system (GPS) device or smartphone is a great way to prevent getting lost, but remember that more remote areas around the lake may have little to no cellular signal. And if you plan to be near or in water, be sure to put your electronics in a sealed plastic bag, just in case.

Collecting Etiquette

Too often, popular collecting sites are closed by landowners or local governments due to litter, trespassing, and vandalism. In many of these cases, the landowners may have been kind enough to allow collectors onto their land, but when people would rather trespass than to simply ask for permission, then we all lose. When collecting, never go onto private property unless you've obtained prior permission, and be courteous; don't dig indiscriminately and don't take more than you need. And by sharing specimens, information, and your enthusiasm, you're likely to be invited back. To ensure great collecting sites for future rock hounds, dig carefully and leave the location cleaner than you found it.

DANGEROUS MINERALS AND PROTECTED ARTIFACTS OF LAKE SUPERIOR

Potentially Hazardous Minerals

The vast majority of minerals in the Lake Superior region are very safe to handle and collect. But a handful pose a health risk under certain conditions. Potentially hazardous minerals included in this book are identified with the symbol shown above, of which there are only a few examples:

- Amphibole group (page 89)—a few varieties are asbestos; asbestiform minerals form as delicate, flexible fiber-like crystals that can become airborne and inhaled, posing a cancer risk; these varieties are rare in this region

- Algodonite and Domeykite (page 87)—contain arsenic; wash your hands after handling

With the above minerals, the primary threat to your health comes with cutting and polishing them, or any other activity that creates dust. Inhaling dust produced by these and any other minerals can be very detrimental to your lungs and respiratory health, potentially introducing toxic particles into your body and can eventually cause cancer. Thankfully, dust inhalation is easily avoided by wearing the proper mask and eye protection.

Artifacts

The Lake Superior region had been occupied by indigenous peoples for thousands of years before settlers arrived. As a result, countless artifacts such as arrowheads and grinding stones have been found throughout the area. But it is important to remember that it is illegal in all cases to disturb or collect these artifacts. They may hold considerable scientific and cultural value, and all finds should be left in place, photographed/recorded and reported to the Bureau of Land Management. Any collecting or tampering with artifacts may incur fines or other penalties.

For more information, contact the Bureau of Land Management's Northeastern States Field Office, which has jurisdiction over Minnesota, Wisconsin, and Michigan, by calling 414-297-4400, or by visiting www.blm.gov.

In Ontario, to report artifacts or other significant discoveries, contact the Ontario Ministry of Heritage by calling 1-888-997-9015, or by visiting www.mtc.gov.on.ca. The Ontario Archaeological Society is another option: www.ontarioarchaeology.org

HARDNESS AND STREAK

There are two important techniques everyone wishing to identify minerals should know: hardness and streak tests. All minerals will yield results in both tests, which makes these tests indispensable to collectors.

The measure of how resistant a mineral is to abrasion is called hardness. The most common hardness scale, called the Mohs Hardness Scale, ranges from 1 to 10, with 10 being the hardest. An example of a mineral with a hardness of 1 is talc; it is a chalky mineral that can easily be scratched by your fingernail. An example of a mineral with a hardness of 10 is diamond, which is the hardest naturally occurring substance on Earth and will scratch every other mineral. Most minerals fall somewhere in the range of 2 to 7 on the Mohs Hardness Scale, so learning how to perform a hardness test (also known as a scratch test) is critical. Common tools used in a hardness test include your fingernail, a U.S. nickel (coin), a piece of glass and a steel pocket knife. There are also hardness kits you can purchase that have a tool of each hardness.

To perform a scratch test, you simply scratch a mineral with a tool of a known hardness—for example, we know a typical steel knife has a hardness of about 5.5. If the mineral is not scratched, you will then move to a tool of greater hardness until the mineral is scratched. If a tool that is 6.5 in hardness scratches your specimen, but a 5.5 did not, you can conclude that your mineral is a 6 in hardness. Two tips to consider: As you will be putting a scratch on the specimen, perform the test on the backside of the piece (or, better yet, on a lower-quality specimen of the same mineral), and start with tools softer in hardness and work your way up. On page 22, you'll find a chart that shows which tools will scratch a mineral of a particular hardness.

The second test every amateur geologist and rock collector should know is streak. When a mineral is crushed or powdered, it will have a distinct color—this color is the same as the streak color. When a mineral is rubbed along a streak plate, it will leave behind a powdery stripe of color, called the streak. This is an important test to perform because sometimes the streak color will differ greatly from the mineral itself. Hematite, for example, is a dark, metallic and gray mineral, yet its streak is a rusty red color. Streak plates are sold in some rock and mineral shops, but if you cannot find one, a simple unglazed piece of porcelain from a hardware store will work. But there are two things to remember about streak tests: If the mineral is harder than the streak plate, it will not produce a streak and will instead scratch the plate itself. Secondly, don't bother testing rocks for streak; they are made up of many different minerals and won't produce a consistent color.

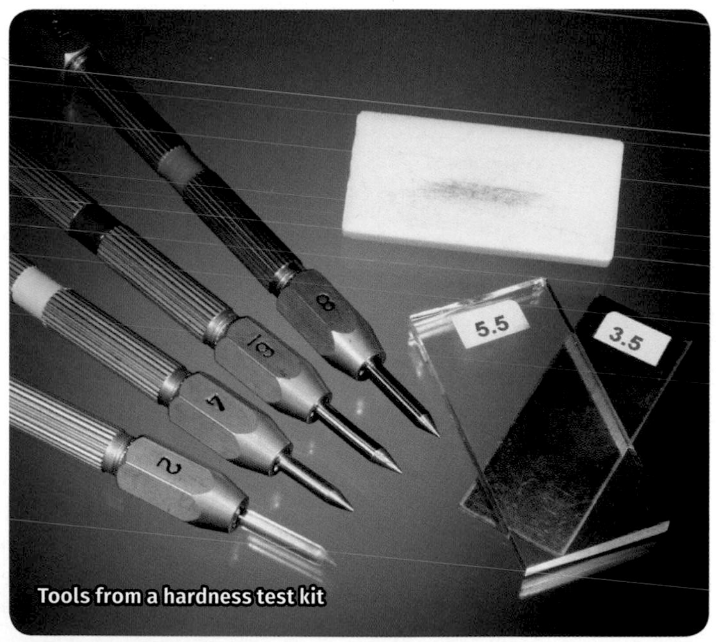

Tools from a hardness test kit

THE MOHS HARDNESS SCALE

The Mohs Hardness Scale is the most common measure of mineral hardness. This scale ranges from 1 to 10, from softest to hardest. Ten minerals commonly associated with the scale are listed here, as well as some common tools used to determine a mineral's hardness. If a mineral is scratched by a tool of a known hardness, then you know it is softer than that tool.

Hardness	Example Mineral	Tool
1	Talc	
2	Gypsum	
2.5		Fingernail
3	Calcite	
3.5		U.S. nickel, brass
4	Fluorite	
5	Apatite	
5.5		Glass, steel knife
6	Orthoclase feldspar	
6.5		Streak plate
7	Quartz	
7.5		Hardened steel file
8	Topaz	
9	Corundum	
9.5		Silicon carbide
10	Diamond	

For example, if a mineral is scratched by a U.S. nickel (coin) but not by your fingernail, you can conclude that its hardness is 3, equal to that of calcite. If a mineral is harder than 6.5, or the hardness of a streak plate, it will instead scratch the streak plate itself (unless impure or weathered to a softer state).

QUICK IDENTIFICATION GUIDE

Use this quick identification guide to help you determine which rock or mineral you may have found. Listed here are the primary color groups followed by some basic characteristics of the rocks and minerals of Lake Superior, as well as the page number where you can read more. While the most common traits for each rock or mineral are listed here, be aware that your specimen may differ greatly.

WHITE OR COLORLESS

If white or colorless and...	then try...
Abundant, soft veins or masses in rocks, often in blocky shapes or steeply pointed six-sided crystals	calcite, page 99
Chalky, soft nodules with a spiderweb- or cauliflower-like exterior texture	datolite, page 133
Small, hard, brittle translucent fragments with a "frosted" appearance and even, uniform thickness	junk (glass), page 167
Abundant, very hard, translucent six-sided crystals or masses within cavities or as rounded pebbles	quartz, page 197
Very hard rock that resembles quartz but has a grainy texture when broken	quartzite, page 203
Small, soft, translucent crystals shaped like faceted balls, grown clustered together	zeolite group (analcime), page 227

Quick Identification Guide (continued)

GRAY OR BLACK

If gray or black and...	then try...
Hard grains or rectangular masses, often with a lustrous silky sheen, found embedded in coarse-grained rocks	amphibole group, page 89
Glassy, coarse-grained, greenish gray, light-colored rock found rarely on Minnesota's shores	anorthosite, page 91
Very abundant, dark, fine-grained rock often containing gas bubbles lined with minerals	basalt, page 97
Very hard, common, opaque masses with waxy surfaces and very sharp edges when broken	chert, page 109
Dark gray, dense rock containing fine to medium grains, often with light-colored flecks or spots	diabase, page 135
Dark, greenish black, coarse-grained rock found primarily on Minnesota's shores	gabbro, page 149
Fairly soft, dense, medium-gray rock composed of tightly packed tiny grains of varying size	graywacke, page 159
Heavy, glassy, very dark, opaque material with countless round bubbles, often with rusty surfaces	junk, page 167
Small, brightly lustrous, flaky crystals, usually embedded in coarse-grained rocks like granite	mica, page 179

GRAY OR BLACK

(continued) **If gray or black and...**	**then try...**
Fairly soft, dense, medium-gray rocks containing perfectly round holes or pockets that seem "out of place"	omarolluks, page 159
Fairly hard, glassy, elongated masses or blocky crystals embedded in dark rocks, especially gabbro	pyroxene group, page 195
Dull, dark gray veins or masses within quartz or rock that are malleable and reveal silvery metal when scratched	silver, page 213
Dense, heavy, dark gray, sometimes layered rock that is very hard and sticks to a magnet	taconite, page 217
Opaque dark brownish-black masses or crusts found on or near copper	tenorite, page 219

TAN OR BROWN

If tan or brown and...	**then try...**
Flat, thin, plate-like crystals that are very brittle and heavy for their size, usually opaque but may be translucent	baryte, page 95
Very hard, translucent, waxy masses, often in ball-like shapes; forms very sharp edges when broken	chalcedony, page 103

Quick Identification Guide (continued)

TAN OR BROWN

(continued)	**If tan or brown and...**	**then try...**
	Soft, gritty, crumbly material that becomes sticky when wet; may form odd, globular masses	clay, page 115
	Hard, uniquely round or blob-like masses, usually found in or near sedimentary rocks along rivers	concretions, page 117
	Soft nodules with a rough or cauliflower-like exterior, usually with a chalky, lighter colored interior	datolite, page 133
	Very common, hard, dull to glassy blocks or masses found primarily within coarse-grained rocks	feldspar, page 139
	Soft, light-colored rocks containing odd patterns resembling gauze, seashells, or snail shells	fossils, page 145
	Soft, chalky rock that can be easily scratched with a pocket knife and fizzes in vinegar	limestone, page 171
	Hard, often layered rocks exhibiting traits typical of sedimentary rocks but of higher hardness or density	metasedimentary rocks, page 177
	Extremely fine-grained rocks that are dense, gritty, and generally soft; may exhibit some layering	mudstone or siltstone, page 181
	Very hard rock that resembles quartz but has a grainy texture when broken	quartzite, page 203

TAN OR BROWN

(continued) **If tan or brown and...**	**then try...**
Light brown to reddish brown, hard, fine-grained rock; may contain colored bands or hollow gas bubbles	rhyolite, page 205
Rough, gritty, sometimes crumbly rock composed of tiny sand grains cemented together	sandstone, page 207
Soft, fine-grained rock consisting of flat, parallel layers that can be separated with a knife	shale, page 211

GREEN OR BLUE

If green or blue and...	**then try...**
Very soft, dark-green masses, usually as crusts or linings within cavities in dark rocks	chlorite group, page 111
Soft, often chalky, light blue masses or crusts alongside copper and/or green malachite	chrysocolla, page 113
Hard, yellow-green, pistachio-colored masses or crusts; less commonly as elongated, grooved crystals	epidote, page 137
Soft, glassy, translucent crystals or veins that break in angular shapes	fluorite, page 143
Tiny, very rare, elongated, vivid blue crystals, usually embedded within calcite	kinoite, page 169

Quick Identification Guide *(continued)*

GREEN OR BLUE

(continued) If green or blue and...	then try...
Soft, vivid green, chalky crusts or fibrous masses grown on or with copper or chalcopyrite	malachite, page 175
Very soft, blue-green, dull masses or linings within cavities in rock	mica group, page 179
Hard, glassy, pale green masses or rounded clusters of short bladed crystals, often with calcite	prehnite, page 187
Fairly soft, opaque green nodules, usually in basalt, with a fibrous or spiderweb pattern; in Michigan	pumpellyite (greenstone), page 191
Soft, translucent, bluish green waxy or glassy masses, most often found in Minnesota	zeolite group (lintonite), page 231

PURPLE OR PINK

If purple or pink and...	then try...
Very hard, glassy, translucent purple masses or crystals	amethyst, page 201
Soft, glassy, translucent crystals, crusts or veins that break in angular shapes	fluorite, page 143
Very soft, crumbly and splintery pink or salmon-colored crystals with an elongated shape, often with calcite	zeolite group, page 223

RED OR ORANGE

If red or orange and...	then try...
Very hard, translucent, waxy masses, often with mottled coloration; forms very sharp edges when broken	chalcedony, page 103
Fairly soft, dull to glassy or nearly metallic red to orange-red crusts or masses on or with copper	cuprite, page 131
Hard, typically opaque, angular crystals or linings inside cavities within rocks, especially basalt	feldspar group, page 139
Very hard, rounded, ball-like crystals embedded in rocks like granite, gneiss, or schist	garnet group, page 151
Soft, dusty, rusty red coatings or masses that may reveal a metallic black color when scratched or broken	hematite, page 161
Very hard, dense, opaque, colorful masses with rough texture and sharp edges when broken, but smooth and waxy when weathered	jasper, page 165
Light reddish brown, hard, fine-grained rock; may contain colored bands or hollow gas bubbles	rhyolite, page 205
Soft, very small, delicate plate-like crystals clustered together in cavities; sometimes has a wheat-sheaf shape	zeolite group (stilbite), page 233

Quick Identification Guide (continued)

YELLOW

If yellow and...	then try...
Very hard, rounded, ball-like crystals in cavities within dark rocks, often alongside epidote	garnet group, page 151
Opaque, chalky, rust-colored coatings or crusts atop other minerals, especially metallic minerals	limonite, page 155
Hard, glassy, greenish yellow, translucent grains or masses within dark rocks like gabbro	olivine group, page 183

METALLIC

If metallic and...	then try...
Rare, brittle veins, often in quartz, usually silvery to brassy in color and may have an iridescent surface	algodonite or domeykite, page 87
Soft, brittle, dark colored veins or masses; scarce and only found in copper-rich regions	chalcocite, page 105
Fairly soft, brightly lustrous, brassy masses or veins embedded in rock, sometimes with green malachite	chalcopyrite, page 107
Soft, malleable reddish orange metal; usually coated in green, blue, red, or black mineral crusts	copper, page 121
Dark brown, rusty crusts or masses, sometimes with a fibrous cross-section; often coated in limonite	goethite, page 155

(continued) **If metallic and...**	**then try...**
Soft, malleable masses consisting of both copper and silver	copper-silver combination, page 129
Abundant, fairly hard mineral that is dark gray when fresh, but usually has a rusty red coating when weathered	hematite, page 161
Tiny, black, highly lustrous and slightly magnetic grains in dark rocks, especially gabbro	ilmenite, page 163
Very soft, lightweight globular masses found on shores	junk (aluminum), page 167
Black, fairly hard masses or grains that bond strongly with a magnet; most often found in dark rocks	magnetite, page 173
Hard, brassy yellow to brownish cube-shaped crystals, masses, or veins; often found in sedimentary rocks	pyrite, page 193
Soft, malleable silvery veins or masses within rock or quartz, usually coated in a dull gray surface material	silver, page 213

Quick Identification Guide (continued)

If multicolored or banded and...	then try...
Very hard, waxy, ball-like masses containing ring-like banding within	agates, page 39
Very hard, rounded masses containing parallel layering (often below some ring-like banding) within	gravitationally banded agate, page 57
Very hard, dense, opaque masses that have a similar appearance to both agates and jasper	jasp-agate, page 61
Very hard masses containing ring-like agate banding as well as ample mineral growths that resemble moss	moss agate, page 65
Very hard masses containing ring-like agate banding as well as circular, radiating structures	sagenitic agate, page 71
Very hard masses containing some tan or gray agate banding and lots of opaque, white, "crackled" quartz	skip-an-atom agate, page 73
Very hard veins of lace-like banded structures, often with lots of calcite, found near Thunder Bay, Ontario	Thunder Bay agate, page 75
Round, dense rocks containing pockets of very hard, reddish agate banding within; found in Minnesota	thunder egg, page 77
Very hard, waxy, rounded masses containing agate banding and tube-like structures throughout	tube or stalk agate, page 79

(continued) **If multicolored and banded and... then try...**

If multicolored and banded and...	then try...
Very hard, dense rock consisting of parallel layers of red jasper and metallic hematite or magnetite	banded iron formation, page 93
Rock that appears to consist of smaller rounded or angular rock fragments cemented together	conglomerate or breccia, page 119
Quite hard, dense rocks containing somewhat coarse mineral grains that run in a generally parallel direction	gneiss, page 153
Abundant coarse-grained rocks with mottled coloration, generally with light and dark colored spots	granite and granitoids, page 157
Very hard, opaque masses with sharp edges when broken, but smooth and waxy when weathered	jasper, page 165
Material that consists of tightly packed rock fragments locked together by a very hard, fine-grained cement	junk (concrete), page 167
Tightly layered, tough rocks that appear similar to sedimentary rocks but are harder	metasedimentary rocks, page 177
Extremely fine-grained rocks that are dense, gritty, and generally soft with near-parallel layering	mudstone or siltstone, page 181

Quick Identification Guide *(continued)*

MULTICOLORED OR BANDED

(continued)	**If multicolored and banded and...**	**then try...**
	Rocks containing large, well-formed blocky crystals suspended in an otherwise finer grained mass	porphyry, page 185
	Coarse, very hard rock that is predominantly white but contains angular red and brown fragments	puddingstone, page 189
	Rough, gritty, sometimes crumbly rock composed of tiny sand grains cemented together	sandstone, page 207
	Dense rock with tight layers of distinct minerals, often with lustrous, "glittery" mica flakes and pyrite	schist, page 209
	Jasper or chert containing layers arranged in wavy or rounded, arching shapes	stromatolites, page 147
	Gray speckled coarse-grained rock, often with orange feldspars; some grains glow under UV light	syenite, page 215
	Green and orange or pink coarse-grained rock resembling granite	unakite, page 221
	Fairly soft radial arrangements of needle-like crystals, often with circular bands, within vesicles	zeolite group (thomsonite), page 235

A Final Note About Rock and Mineral Identification

When using this book to identify your rock and mineral discoveries, always remember that your specimens can (and likely will) differ greatly than those pictured. Rock and mineral identification isn't always easy, and when a specimen is weathered or altered by external forces, it can appear completely different than it "should." The photos in this book are meant as a general guide; learning the key characteristics of each rock or mineral and which traits are constant, such as hardness and crystal shape, will bring you the most success. With a basic understanding of quartz, for example, you'll be able to identify even the most poorly formed specimens.

Many of the Lake Superior region's rocks are also incredibly ancient. Eons of erosion from wind, waves, ice, and chemicals have dramatically changed the appearance of many rocks, which can make identification difficult. Take this specimen of a rock called gabbro, for example. Externally, it appears somewhat light-colored with what look like dots of rust all over it. But upon breaking it in half, you can see that the unweathered interior is darker and more crystalline in appearance. The interior reveals how typical gabbro should look, but if all you had to go by was the exterior, you'd likely have trouble identifying this common rock.

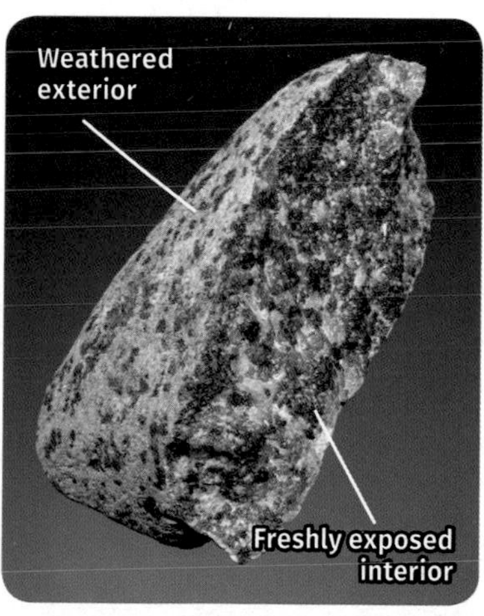

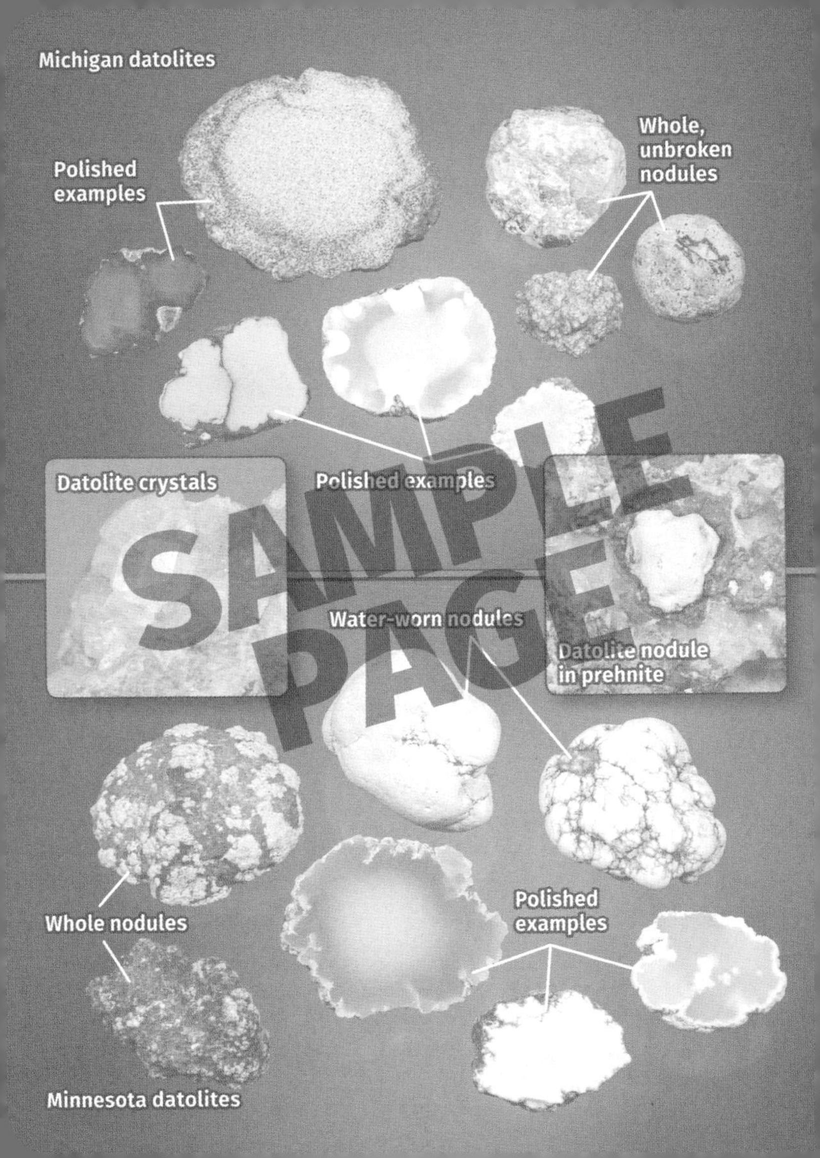

Michigan datolites

Polished examples

Whole, unbroken nodules

Datolite crystals

Polished examples

Water-worn nodules

Datolite nodule in prehnite

Whole nodules

Polished examples

Minnesota datolites

Sample page

Hardness: From 1 to 10 **Streak:** Color

Primary Occurrence

Environment: A generalized indication of the types of places where this rock or mineral can commonly be found. For the purposes of this book, the primary environments listed are **shores** (both of Lake Superior and nearby inland lakes), **rivers** (encompassing riverbeds and riverbanks), **pits** (gravel pits, quarries, or other dig sites where earth has been removed), **outcrops** (any exposed bodies of rock both natural and manmade, such as where road construction has cut through a rock formation), and **mines** (primarily dumps, or waste-rock piles left over at mine sites).

What to Look for: Common and characteristic identifying traits of the rock or mineral.

Size: The general size range of the rock or mineral. The listed sizes apply more to minerals and their crystals than to rocks, which typically form as enormous masses.

Color: The colors the rock or mineral commonly exhibits.

Occurrence: The relative difficulty of finding this rock or mineral. "**Very common**" means the material is abundant and takes no effort to find if you're in the right environment. "**Common**" means the material can be found with little effort. "**Uncommon**" means the material may take a good deal of hunting to find; most minerals fall in this category. "**Rare**" means the material will take great lengths of research, time, and energy to find, and "**very rare**" means the material is so scarce that you may likely never find it.

Notes: These are additional notes about the rock or mineral, including how it forms, how to identify it, how to distinguish it from similar minerals, and interesting facts about it.

Where to Look: Here you'll find specific regions or towns where you should begin your search for the rock or mineral.

olished agate
Quartz-filled center
Agate in basalt
Waxy luster
Limonite surface coating (yellow-brown)
Typical small agate "chips" weathered from larger agates
Common beach-worn agates

Primary Occurrence

Agates, general

Hardness: 6.5–7 **Streak:** White

Environment: All environments

What to Look for: Very hard, translucent, red or brown rounded masses of waxy material containing ring-like bands within

Size: Agates range from tiny fragments to finds rarely larger than a fist

Color: Multicolored; varies greatly, but banding is primarily red, brown, yellow, white to gray, often with colorless layers

Occurrence: Uncommon

Notes: Without question, Lake Superior's most famed and popular collectibles are agates. These gemstones consist primarily of chalcedony (page 103), which is a microcrystalline variety of quartz (page 197), and they grew within cavities in rocks, most often within the vesicles (gas bubbles) in volcanic rocks. This makes them generally more-or-less ball-shaped (such formations are called nodules). Their primary defining feature is, of course, their concentric banding, but like the layers of an onion, the bands aren't revealed until weathering (usually the glaciers) has broken open the nodules. Jasper (page 165), chert (page 109), chalcedony and agates—all of which consist of microcrystalline quartz—all exhibit conchoidal fracturing (when struck, circular cracks appear), extreme hardness, and a waxy luster when found as weathered pebbles, but jasper and chert are not translucent like chalcedony and agates, and common chalcedony lacks agates' bulls-eye banding. While jasper and chert can show layering as well, their layers are usually thick, flat, parallel, and very opaque. Whole agate nodules are scarce but usually show a waxy, pockmarked surface; far more common are small, glossy, carnelian-colored fragments.

Where to Look: Opportunities to find agates are plentiful; just look where the average tourist doesn't—muddy river banks, remote beaches, and inland gravel pits (with permission).

Rough, natural adhesional banded agates with choice patterns

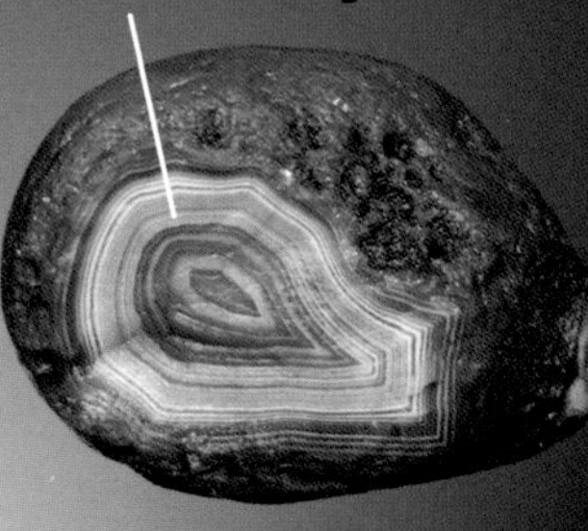

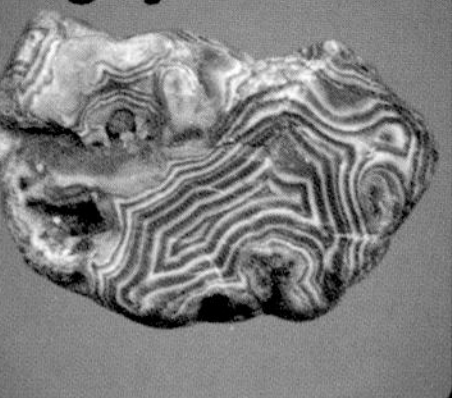

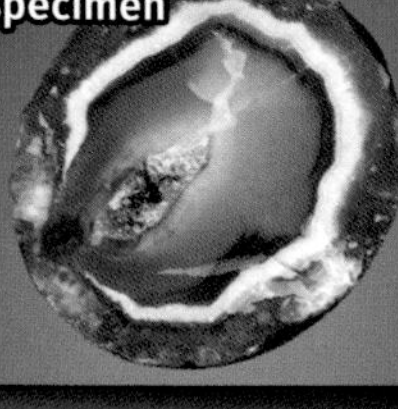

Polished adhesional banded agates

Primary Occurrence

Agates, adhesional banded

Hardness: 6.5–7 **Streak:** White

Environment: All environments

What to Look for: Very hard, translucent, red or brown rounded masses of waxy material containing ring-like bands within

Size: Agates range from tiny fragments to finds rarely larger than a fist

Color: Multicolored; varies greatly, but banding is primarily red, brown, yellow, white to gray, often with colorless layers

Occurrence: Uncommon

Notes: While there are many varieties of Lake Superior agates present in the region, adhesional banded agates are among the most common and most popular. These are the quintessential agates: often perfect band-within-a-band patterns in which each inner band is adhered to and mimics the shape of the outer ones. Known among collectors as fortification agates, due to their resemblance to the concentric walls of a fort or castle, examples with bold banding of alternating color can be among the most valuable of agates. While we understand what agates are mineralogically, how they form is still somewhat of a mystery; while there are many theories, none is conclusive. But however their repeating layers form, adhesional banded agates represent an uninterrupted layering process and are considered "ideal" agates. Sometimes, due to a drop in the amount of silica (quartz material) available to a developing agate, the center of a specimen may consist of larger quartz crystals, rather than agate banding.

Where to Look: Some of the best and most valuable specimens have originated along Minnesota's shores, especially at inland gravel pits in the Duluth area. The northwestern shore of the Keweenaw Peninsula is lucrative as well, as is the shore from Little Girls Point westward into Wisconsin. In Ontario, the Thunder Bay area shores are worth a look.

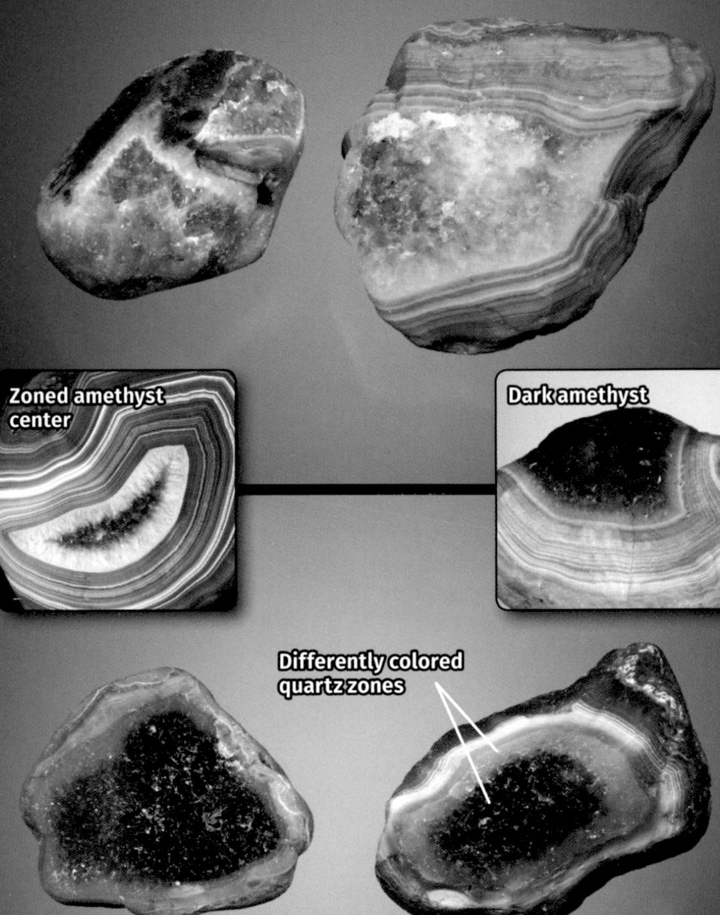
Rough agates with amethyst (purple quartz) centers
Zoned amethyst center
Dark amethyst
Differently colored quartz zones
Rough agates with smoky quartz (dark gray to black quartz) centers

Agates, colored quartz

Hardness: 6.5–7 **Streak:** White

Primary Occurrence

Environment: Shores, rivers, pits, outcrops

What to Look for: Very hard, translucent, rounded masses containing ring-like bands and purple or black quartz within

Size: Most clear examples will be an inch or two, rarely larger

Color: Agate banding varies and is primarily red, brown, white; quartz centers are gray-brown to black, purple, rarely green

Occurrence: Uncommon to rare

Notes: Agates with a large central mass of coarsely crystallized white quartz are quite common. Since agates form their layers from the outside inward, we know that the central quartz was last to form. Chalcedony banding requires more silica (quartz material) to form than typical quartz, so these examples represent agates that used up most of the available silica when forming their banding, leaving a depleted amount that could only form coarse quartz at the center. These agates are generally less desirable by collectors—unless the quartz is colored. Purple amethyst, gray smoky quartz, brown "root beer quartz," and very rare greenish quartz are all present within agates of the region, and all are caused by chemical impurities trapped within the microscopic spaces inside the quartz's crystal structure. Coloration can vary in intensity as well as in consistency; some examples show color zoning, in which white quartz turns to colored quartz partway through. Note, however, that agates with reddish or yellowish quartz centers are generally just stained by iron impurities; the color in these agates is found in the cracks between the crystals rather than inside them.

Where to Look: Agates with colored quartz don't generally appear in any one area with more frequency than others. But areas with higher concentrations of agates will be more lucrative. Gravel pits, beaches, and rivers in the Duluth, Minnesota, area have long been excellent sources.

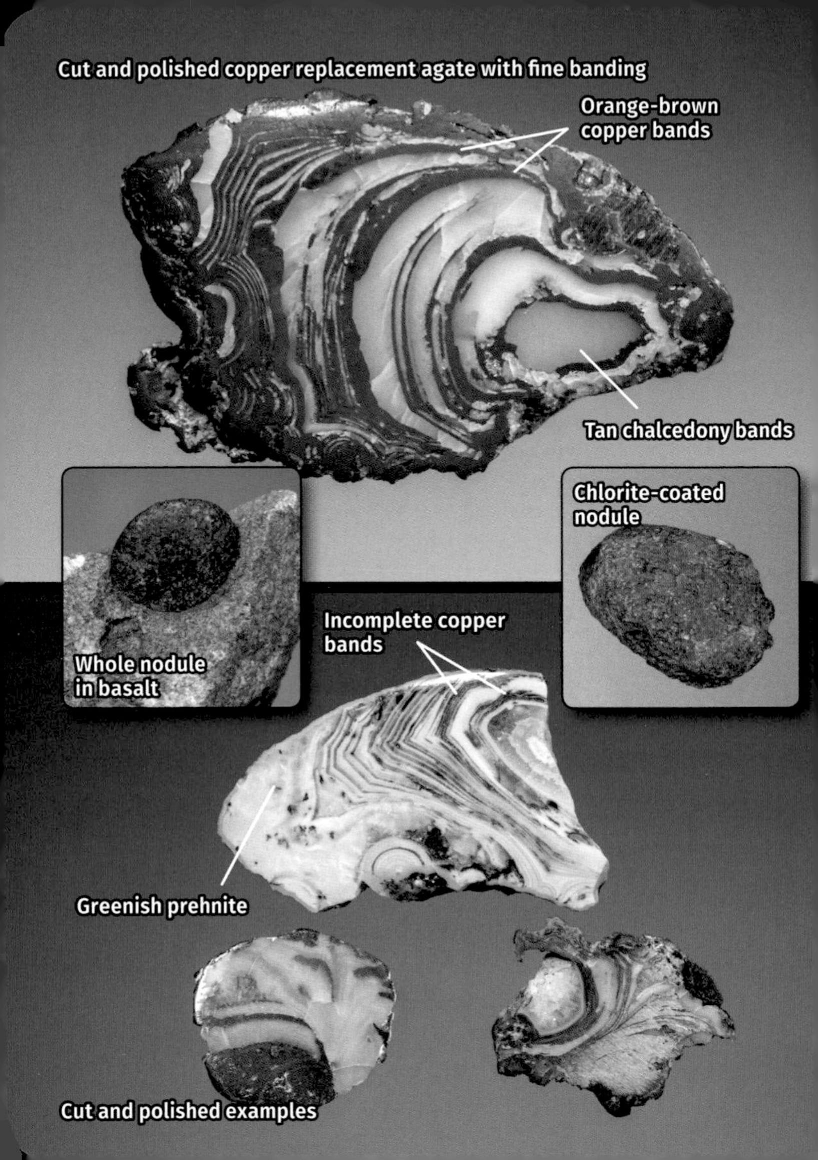

Cut and polished copper replacement agate with fine banding

Orange-brown copper bands

Tan chalcedony bands

Whole nodule in basalt

Chlorite-coated nodule

Incomplete copper bands

Greenish prehnite

Cut and polished examples

Primary Occurrence

Agates, copper replacement

Hardness: N/A **Streak:** N/A

Environment: Mines

What to Look for: Small, dark nodules embedded in basalt that contain both agate banding and copper within

Size: Always small; specimens larger than an inch are rare

Color: Dark gray-green exteriors; agate banding is usually white to cream-colored or tan; copper is metallic red when cut

Occurrence: Very rare in Michigan; not found elsewhere

Notes: Michigan's copper replacement agates are among the rarest, most collectible, and most enigmatic of all Lake Superior agates. Called "copper agates," for short, these small agates actually contain native copper, frequently as little masses or flecks throughout, but also as agate-like bands, alternating with white or tan chalcedony layers. Agate formation is still poorly understood, but the unique and anomalous formation of copper agates only confounds the issue. It is believed, however, that the copper has replaced specific bands—likely very impure chalcedony bands that dissolved in hot groundwater, leaving band-shaped voids in which copper was later deposited—hence their name. They are generally thumbnail-size or smaller, often found as chlorite-coated nodules still embedded in their host basalt, and can also include intergrown calcite and prehnite. They are only found at certain mine dumps and must be broken free from their host rock, then carefully sawn open to see if they contain copper—many do not. But with research, patience, and great effort, these rarities can still be found.

Where to Look: A few of the copper mine dumps have produced copper agates. Dump piles and old mine sites around Kearsarge and Calumet, north of Houghton, are particularly well-known, but other locations may exist as well.

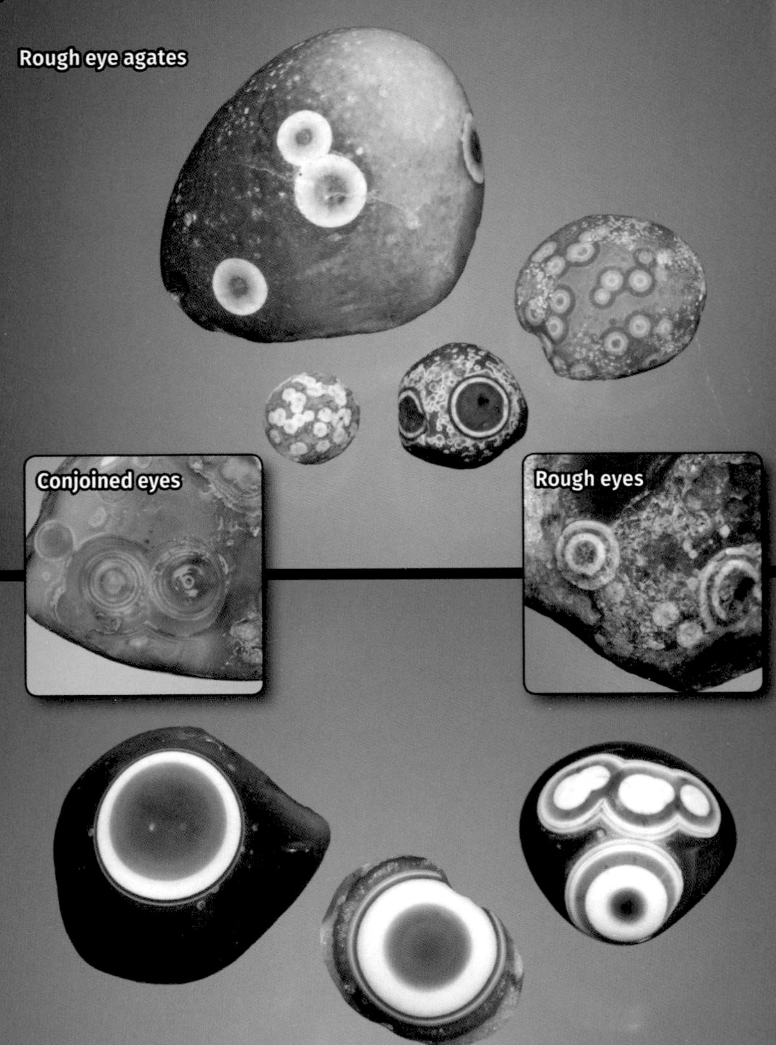
Rough eye agates
Conjoined eyes
Rough eyes
Polished eye agates

Agates, eye

Hardness: 6.5–7 **Streak:** White

Primary Occurrence

Environment: All environments

What to Look for: Very hard, translucent, red or brown rounded masses of waxy material with eye-like surface spots

Size: Eye agates are generally smaller than a golf ball

Color: Multicolored; varies greatly, but banding is primarily red, brown, yellow, white to gray, often with colorless layers

Occurrence: Uncommon

Notes: Eye agates, also called "fish eyes," are among the most endearing and popular varieties of Lake Superior agates. They bear small, perfectly circular banded spots resembling eyes on their outer surfaces. Some finds have only one "eye," but many have clustered groups of them. While they may look like orbs, they are actually hemispheres, or half-spheres, formed on the outer surface of an agate and extending inward, like a shallow bowl. Scientifically, they are enlarged spherulites; spherulites are tiny hemispheres of chalcedony that make up the thickest, outermost layer of an agate. In eye agates, certain spherulites were provided with more silica (quartz material) than they needed and continued to grow in size. This is interesting because eyes are far more common on small agates—sometimes pea-size agates will have the best eyes—indicating that a surplus of available silica will continue to affect a developing agate that has already filled its vesicle. The ends of tubes (page 79) can look like eyes, but tubes are long, cylindrical structures, often with a hollow center.

Where to Look: Eyes tend to be more common in smaller agates; any Minnesota or Wisconsin shoreline will yield little specimens among the beach cobbles. The Keweenaw Peninsula, between Eagle River and Copper Harbor, produces small agate nodules often covered with eyes.

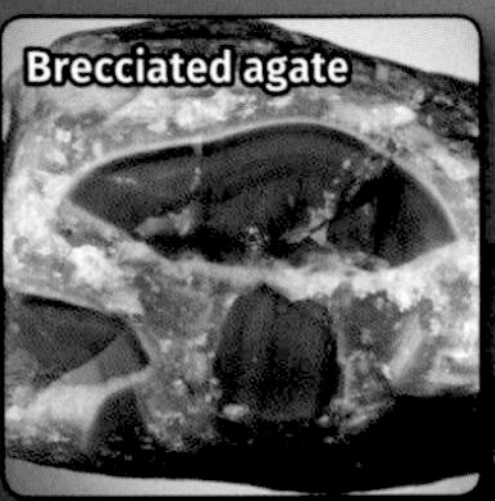

Right inset specimen courtesy of Bob Wright

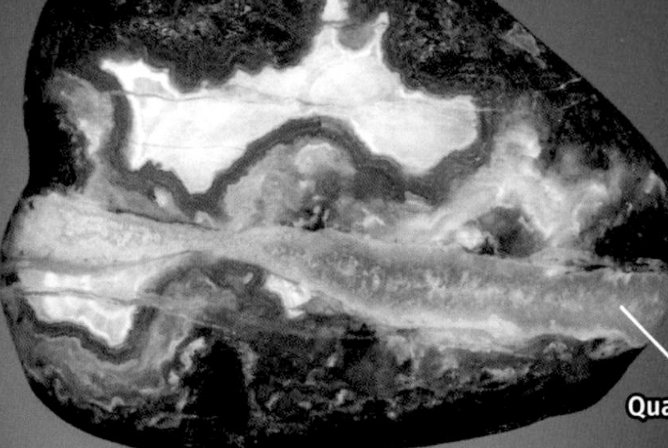

Rough faulted agate

Agates, faulted & brecciated

Hardness: 6.5–7 **Streak:** White

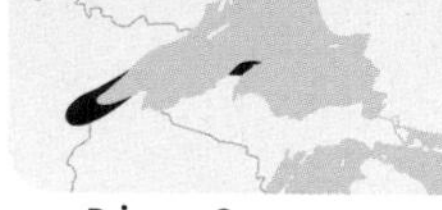

Primary Occurrence

Environment: Shores, rivers, pits

What to Look for: Very hard, waxy, brown masses of material containing bands or layers that appear broken or interrupted

Size: Faults can occur in agates of any size, but they are most often seen in agates measuring a couple inches or more

Color: Multicolored; varies greatly, but banding is primarily red, brown, yellow, white to gray, often with colorless layers

Occurrence: Rare

Notes: Most Lake Superior agates bear signs of their 1.1 billion years of existence—cracks, mineral stains, weathering—but some show more dramatic evidence of their harsh past. Faulted and brecciated agates are examples of agates that were damaged even before they were freed from their host rock. The region's past volcanic activity and associated earthquakes and pressure cracked and broke some nascent agates, fragmenting their banding patterns and misaligning the pieces—sometimes significantly. Later, solutions containing silica (quartz material) flowed into these partially crushed agates and hardened, "healing" the pieces back together in a solid mass. Faults are one result, and they range from finds with slightly misaligned banding segments to those where the agate pattern has large gaps and is separated by masses or veins of quartz. Breccia is the more dramatic result, showing a jumble of completely disjointed banded agate segments "floating" in quartz. Specimens with both phenomena are called "ruin agates."

Where to Look: The Keweenaw Peninsula, especially in the Copper Harbor area, is well-known for agates bearing countless faults. Elsewhere, the usual agate environments can produce rare examples; Minnesota and Wisconsin shorelines and pits, especially in the Duluth and Superior area, are lucrative.

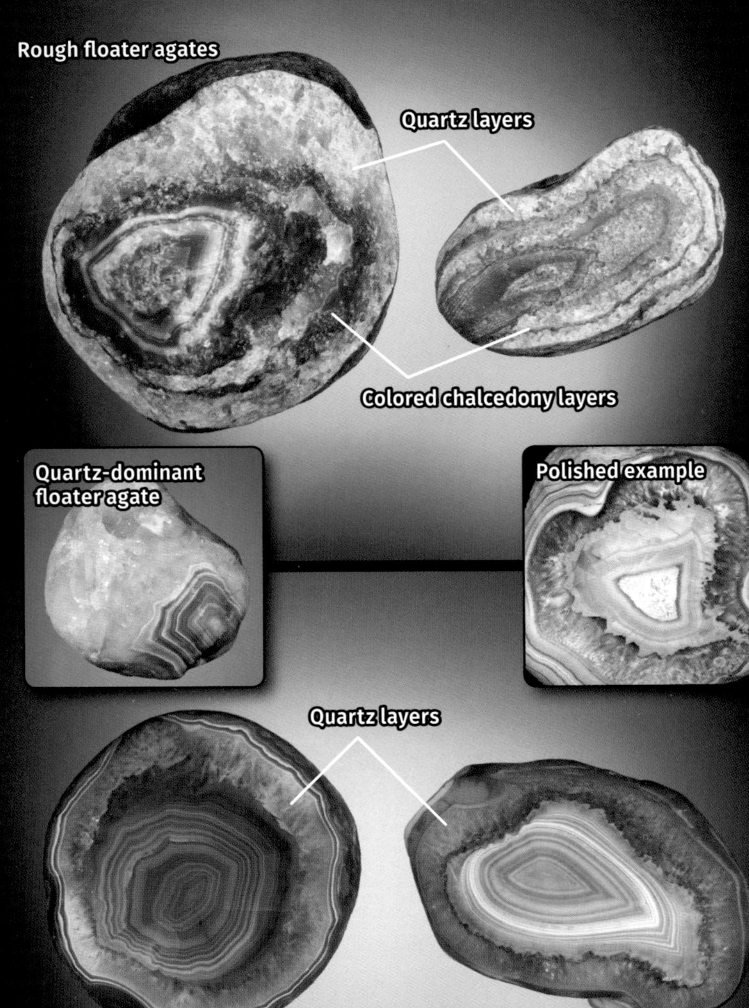
Rough floater agates
Quartz layers
Colored chalcedony layers
Quartz-dominant floater agate
Polished example
Quartz layers
Polished floater agates

Primary Occurrence

Agates, floater

Hardness: 6.5–7 **Streak:** White

Environment: Shores, rivers, pits

What to Look for: Very hard, waxy masses of material containing colored banding as well as layers or non-central masses of quartz

Size: Floater agates are some of the largest agates; specimens larger than a softball are rare but known

Color: Multicolored; varies greatly, but banding is primarily red, brown, yellow, white to gray, often with colorless layers

Occurrence: Uncommon

Notes: It is understood that in order for an agate to form a perfect pattern, as seen in the finest adhesional banded agates (page 41) it needs an uninterrupted, steadily replenished supply of silica (quartz material) until the vesicle is filled with layered chalcedony. But if the silica supply is not constant, instead increasing and decreasing in concentration, a regular pattern will not result. It is believed that this is how floater agates formed; periods of high silica availability formed agate banding, but when the silica supply dipped, only coarsely crystallized quartz could form (quartz requires less silica to form than chalcedony and agates do). In some specimens, this fluctuation only happened once, resulting in a central "island" of banded chalcedony "floating" in a "sea" of quartz, hence their name. But in other examples, it happened multiple times, appearing as alternating chalcedony and thick quartz bands. While viewed as less valuable, floaters are still popular and can contain interesting features that can be seen under the quartz when polished.

Where to Look: Floater agates are one of the most abundant types of agates, found in any typical agate setting, including almost any southern or western Lake Superior beach. Rivers and gravel pits in the Duluth, Minnesota, area are lucrative.

Cut and polished fragmented membrane agate

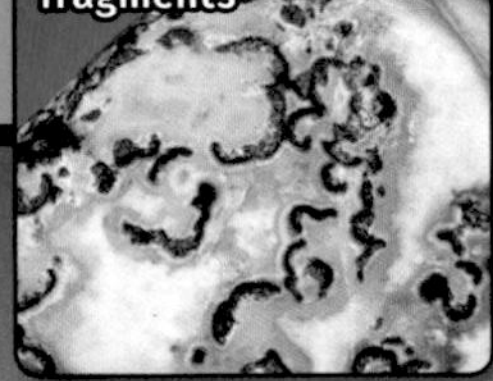

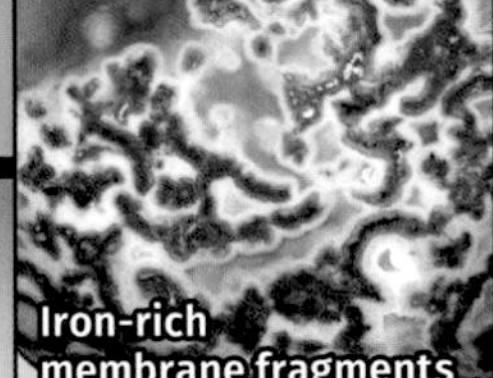

Banded agate center

Membrane fragments

Polished fragmented membrane agate

Agates, fragmented membrane

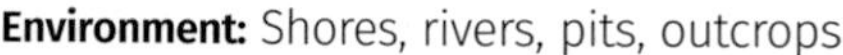

Hardness: 6.5–7 **Streak:** White

Primary Occurrence

Environment: Shores, rivers, pits, outcrops

What to Look for: Very hard, waxy masses of material containing colored banding as well as curved, ribbon-like fragments

Size: These agates can occur in any size, but the characteristic fragments are typically seen in specimens an inch or more

Color: Multicolored; varies greatly, but banding is primarily red, brown, yellow, white to gray, often with colorless layers

Occurrence: Uncommon

Notes: Basalt (page 97) is one of the most prevalent volcanic rocks of the Lake Superior region. It formed when lava was pushed up toward the earth's surface; in the process, gas bubbles, known as vesicles, were trapped within the rock. As the fresh rock cooled and hardened, those gases, along with hot groundwater, began to interact with the new basalt and form minerals within its cavities. Chlorite (page 111) and celadonite (page 179) were common results, grown as thin crusts or membranes that lined the inside of the vesicles. Agates could then form within this soft membrane, which usually weathers away quickly after an agate has been exposed. But sometimes, disturbances during agate formation caused the membrane to break up and sink into a developing agate. Called fragmented membrane agates, they contain curved ribbon-like fragments of the vesicle-lining minerals, often interrupting or separating banded portions. The fragments weather more easily than the agate and often appear pitted and recessed. They also often appear in just one half of an agate (presumably the bottom of the agate as it formed).

Where to Look: Fragmented membrane agates can turn up anywhere; Minnesota's shores and adjoining rivers have produced specimens, as have eastern Michigan's beaches.

Rough agate geodes

Druzy quartz-lined central cavity

Small agate geode

Calcite (white) inside thin agate geode

Large calcite crystal and orange laumontite crystals within agate and quartz geode in basalt

Specimen courtesy of Christopher Cordes

Primary Occurrence

Agates, geode

Hardness: 6.5–7 **Streak:** White

Environment: All environments

What to Look for: Very hard, waxy masses of material containing colored banding as well as a hollow central cavity

Size: Varies greatly; typically golf ball- to softball-size

Color: Multicolored; varies greatly, but banding is primarily red, brown, yellow, white to gray, often with colorless layers

Occurrence: Rare

Notes: Geodes are more or less round formations defined by having a hollow center, often lined with small crystals, and they can be composed of a variety of rocks or minerals, including agate. Since we know that agates form layer by layer from the outside inward and require a steady supply of silica (quartz material) to form, a hollow agate geode indicates an interrupted agate-formation process in which the silica supply dried up and was never replenished. This resulted in an "unfinished" agate with a hollow center, often lined with a druzy coating of small quartz crystals. Sometimes this cavity became the space in which other minerals, such as calcite or zeolites, later formed, making for some of the most unique Lake Superior agates. But agate geodes are generally scarcer than other kinds of agates, as the weight of the glaciers undoubtedly crushed thin ones with relative ease. Thicker-walled agate geodes are therefore more typical, and the hollow center is often quite small. The exception are agate geodes still embedded in their host rock, which can retain many delicate features and crystals.

Where to Look: The Grand Marais, Minnesota, area beaches and outcrops have produced amazing agate geodes containing crystals of other minerals. Elsewhere, they can typically be found anywhere other varieties of agates are found.

ugh gravitationally banded agates
Parallel horizontal layers
Color variation
along bands
Polished specimen
Quartz center
pical adhesional
nds
Horizontal layers
ugh gravitationally banded agate

Agates, gravitationally banded

Hardness: 6.5–7 **Streak:** White

Primary Occurrence

Environment: All environments

What to Look for: Very hard, waxy masses of material containing parallel layering, often with regular agate banding above

Size: Agates range from tiny fragments to finds rarely larger than a fist

Color: Multicolored; varies greatly, but banding is primarily red, brown, yellow, white to gray, often with colorless layers

Occurrence: Uncommon

Notes: Gravitationally banded agates are among the most unique kinds of Lake Superior agates. Better known as water-level agates or onyx, these agates derive their scientific name from the fact that only gravity could have produced the flat, parallel layers of chalcedony seen in these specimens. That makes them quite important for agate research for a variety of reasons; not only did they form completely differently from typical adhesional banded agates (page 41), but we know which direction was "up" when they were forming. It is thought that they developed from silica (quartz material) solutions that were either too watery, preventing it from adhering to the walls of the vesicle (gas bubble), or that contained another chemical that caused the developing chalcedony to clump up and sink. Other puzzling aspects include individual horizontal layers that change color along their length, as well as the phenomenon of some horizontal layers turning into typical adhesional layers at certain portions of the agate. All of these atypical agate features make them among the easiest of varieties to identify.

Where to Look: Minnesota's shores and adjoining rivers are perhaps the best areas in the region to hunt these agates, but they do turn up most anywhere agates are found. Eastern Michigan's shores and Wisconsin gravel pits are also good locales.

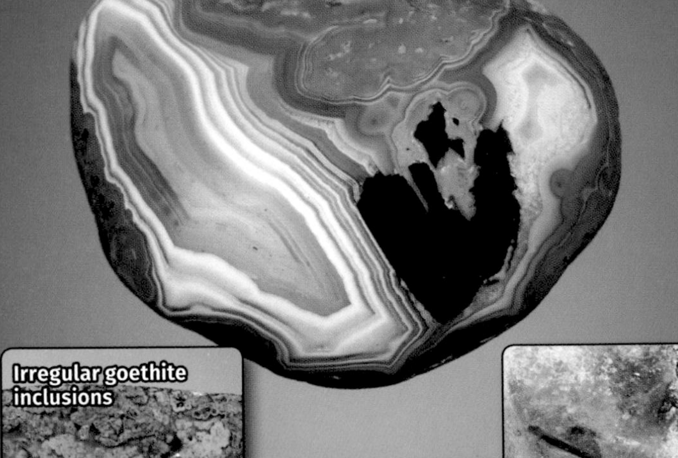

Polished agate with large iron-rich calcite or siderite inclusion (black)

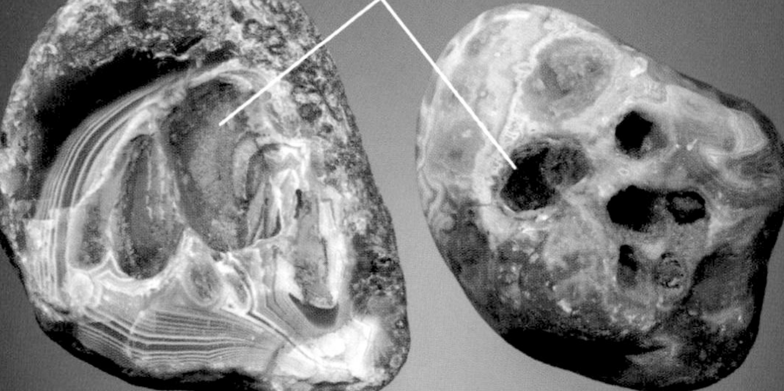

Rough agates with crystal impressions

Agates, inclusions & impressions

Hardness: 6.5–7 **Streak:** White

Primary Occurrence

Environment: All environments

What to Look for: Agates containing what appear to be foreign materials; also, agates that have angular or geometric holes

Size: Inclusions and impressions are found in agates of any size

Color: Multicolored; varies greatly, but banding is primarily red, brown, yellow, white to gray, often with colorless layers

Occurrence: Uncommon

Notes: Agates were rarely the only mineral to form in a vesicle (gas bubble), and they also weren't the first. In addition to the vesicle-lining membrane (visible in fragmented membrane agates, page 53) that formed shortly after the host rock cooled, other minerals like calcite, baryte, goethite and zeolites vied for space in a cavity both before and during the formation of agates. In most cases, the chemicals contributing to agate growth and the ones contributing to the other minerals were incompatible; they did not mix, and the developing agate "pushed aside" the other minerals. The result is agates with embedded mineral inclusions; growths of other minerals, including crystals, like hexagonal points of calcite, or earthy "blobs," like irregular masses of goethite, all of which typically interrupt the agate banding and force it to accommodate them. In some cases, these soft mineral inclusions dissolved after the agate hardened. This left behind hollow voids that have the angular, geometric shapes of the crystals that made them; these are called impressions.

Where to Look: Many agates will show inclusions and impressions. Michigan's famous copper replacement agates contain native copper, found in Keweenaw Peninsula mine dumps, and the Michipicoten area in Ontario produces agates with huge calcite inclusions.

Rough jasp-agates
Poorly defined layers
Polished jasp-agate
Small areas of
agate banding
Opaque coloration
Dark crystal inclusions
Polished jasp-agate

Agates, jasp-agate

Hardness: 6.5–7 **Streak:** White

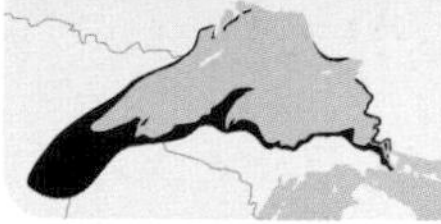

Primary Occurrence

Environment: Shores, rivers, pits, outcrops

What to Look for: Very hard, waxy masses that may be partially banded or layered; specimens may have translucent and opaque parts

Size: Jasp-agates can be some of the largest agate-like material available; while rare, boulder-size masses are known

Color: Usually multicolored; varies greatly, including red to brown or tan, orange to yellow, gray to nearly black

Occurrence: Common

Notes: Quartz can form in a variety of ways, and microcrystalline varieties, which feature tightly bonded microscopic quartz crystals, are very common. Examples include materials like chalcedony (page 103)—the primary material in agates, which consists of tiny fiber-like quartz crystals—and jasper (page 165). Each of these materials consists of silica (quartz material), just formed in different ways. So it should be no surprise that these materials can sometimes occur together in the same masses. Jasp-agates are one possible result; they are a poorly defined variety of agate-like material that usually contains some translucent chalcedony and ample opaque jasper. Very few jasp-agates should actually be labelled "agate," because they simply lack most agate-like qualities, especially concentric banding. Some banding may be present, but it is usually poorly formed and interrupted by opaque masses of jasper, or it is relegated to very small spots and makes up a small percentage of the specimen. Nevertheless, this material is frequently mistaken for true agate by amateurs.

Where to Look: Jasp-agate is considered a low-grade agate material and is generally not widely sought after. They do turn up on lakeshores, but by far the most lucrative sources are Minnesota and Wisconsin gravel pits in the Duluth-Superior area.

Polished Keweenaw paint agate in basalt
Basalt host rock
Faults
Faulted banding
Iron-stained cracks
Whole nodules
Polished nodules

Agates, Keweenaw

Hardness: 6.5–7 **Streak:** White

Primary Occurrence

Environment: Shores, rivers, outcrops

What to Look for: Very hard, rounded masses of waxy material containing ring-like bands within, often with rough exteriors

Size: Varies greatly; most examples are usually golfball-size or smaller, but some measuring a foot or more have very rarely been found

Color: Multicolored; varies greatly, but generally red to orange, white to tan or gray, occasionally bluish

Occurrence: Uncommon; only found in the Keweenaw Peninsula

Notes: While the agates of Michigan's Keweenaw Peninsula aren't typically considered a separate agate variety—most exhibit classic adhesional banding (page 41)—there are enough unique traits present in the peninsula's agates to warrant special mention. The Keweenaw is home to several outcrops that produce agates still embedded in their host rock, typically basalt. This makes round, whole nodules more abundant here than in many other areas. As such, most are relatively intact and exhibit the popular, nearly opaque "paint" coloration (page 67), indicating that they've long been protected from weathering. These whole nodules typically have rough, pock-marked exteriors. Also characteristic of many of the peninsula's agates is dramatic cracking caused by pressure in their shifting host rock. Most of these cracks, often numerous and stained by iron, were later "healed" together by quartz; some are faulted and show bands out of alignment (see page 49).

Where to Look: The Brockway Mountain area, near Copper Harbor, is famous for its whole agate nodules, usually found in or near their host rock. The beaches along the northwestern side of the Keweenaw, between Eagle River and Copper Harbor, are lucrative, and the Keweenaw Point area is home to agates with many faults and cracks.

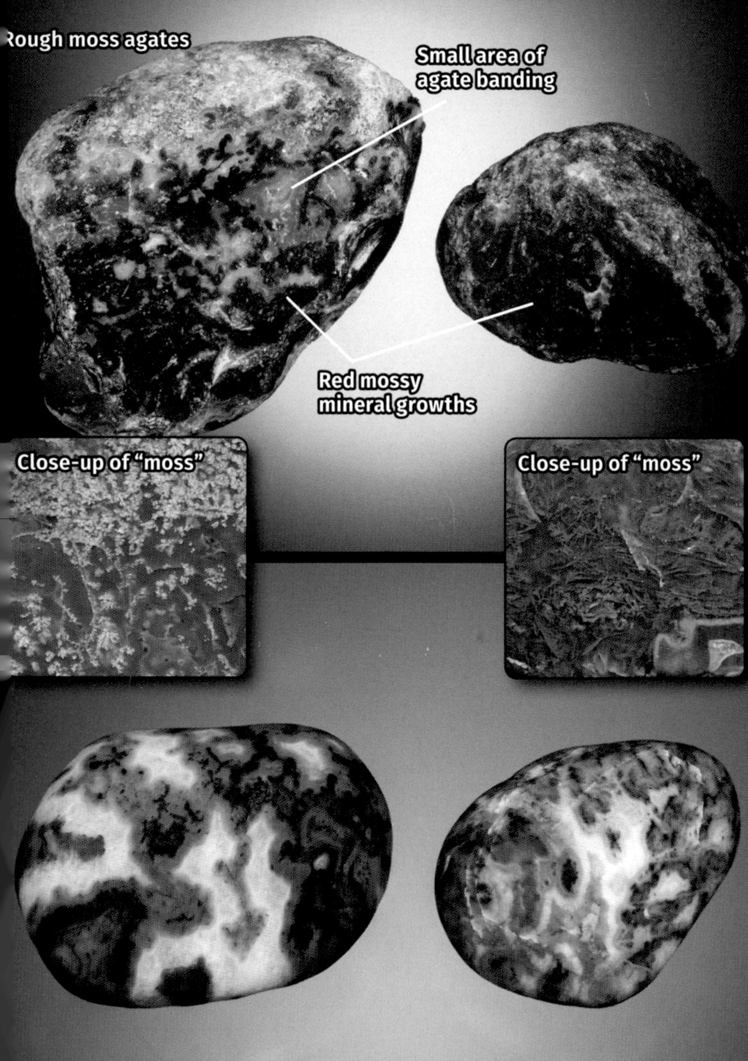
Rough moss agates
Small area of agate banding
Red mossy mineral growths
Close-up of "moss"
Close-up of "moss"
Water-worn agates with mossy inclusions
Specimens courtesy of Bob Wright

Agates, moss

Hardness: 6.5–7 **Streak:** White

Primary Occurrence

Environment: Shores, rivers, pits, outcrops

What to Look for: Very hard, reddish brown waxy masses containing irregular thread- or moss-like patterns within

Size: Moss agates can range widely, from pebbles to boulders

Color: Multicolored; varies greatly, but is primarily red, brown, yellow, green, white to gray, with some colorless areas

Occurrence: Uncommon

Notes: A typical agate may have a few mineral inclusions (page 59), but in moss agates, mineral inclusions are the primary feature. Wispy, thread-like tangles of mineral growths—many of which really do resemble living moss, hence their name—predominate in these specimens. Moss agates formed when a chemical reaction involving iron compounds occurred within a still-soft mass of nascent chalcedony. In many cases, it can be argued that moss agates are not true agates, as most lack any agate-like concentric banding. This is largely because many moss agates did not form in vesicles (gas bubbles) but instead developed within larger cavities or veins that were not conducive to the development of concentric bands. Some examples can, however, contain small portions of agate banding, which formed after the mossy growths and usually sectioned off in little pockets. Because they formed in larger cavities, moss agates can (rarely) be found as boulder-size masses; most stories of "the largest agate ever found" usually refer to a moss agate.

Where to Look: Moss agates are an abundant type; you'll find them most anywhere you can find other agates. Northern Wisconsin's shores produce nicely rounded specimens, and gravel pits in the Duluth, Minnesota, (and southward) area yield many, rarely including huge boulder-size masses.

Polished paint agates

Cut nodule with chlorite exterior

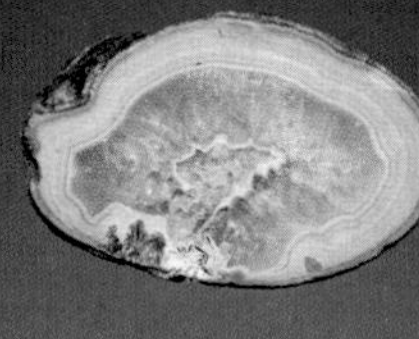

Rough paint agate

Polished paint agates

Agates, paint

Hardness: 6.5–7 **Streak:** White

Primary Occurrence

Environment: All environments

What to Look for: Very hard, waxy rounded masses containing ring-like bands of nearly opaque orange or brown color

Size: Paint agates are rarely larger than your fist

Color: Multicolored; varies, but typically orange to reddish brown, yellowish to tan, pink, rarely green to dark green

Occurrence: Uncommon

Notes: Paint agates, known among collectors simply as "paints," are among the most popular varieties of Lake Superior agates. They are characterized by having dense, nearly opaque coloration and appearing less translucent than typical agates, like their colors were "painted on," hence their name. They are most typically found in shades of orange, tan, and brown (some very rare specimens also contain dark green coloration), and they usually show adhesional- (page 41) or floater-type (page 51) banding structure. The source of the coloration was long thought to be solely the result of higher concentrations of color-causing iron impurities, but we now know that it is actually the result of the agates not yet being significantly weathered. As agates are exposed to the elements, some of their impurities are lost while the coloration of others changes, becoming more translucent. This contrast is observable when we find whole agate nodules still well-protected in their host rock; most of the time, they will exhibit paint coloration. As "fresher" agates, paints may also still have an intact dark outer coating of chlorite.

Where to Look: Paradise Beach, near Grand Marais, Minnesota, is renowned for its stunning paint agates, but much of the area is privately owned so be aware of property lines. Many of the Keweenaw Peninsula's agates are also paint agates.

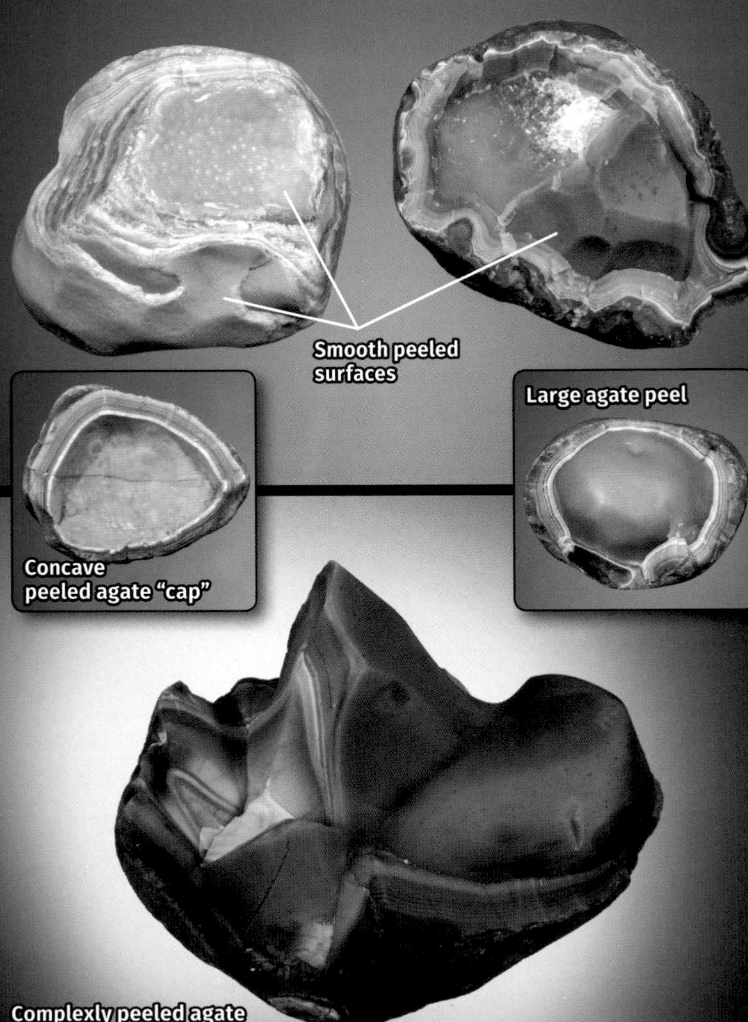
Rough peeled agates
Smooth peeled surfaces
Large agate peel
Concave peeled agate "cap"
Complexly peeled agate

Agates, peeled

Hardness: 6.5–7 **Streak:** White

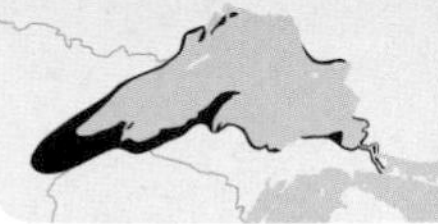

Primary Occurrence

Environment: Shores, rivers, pits

What to Look for: Very hard, waxy rounded masses with ring-like bands and some exceptionally smooth exterior surfaces

Size: Peels are best observed on agates golfball-size and larger

Color: Multicolored; varies greatly, but banding is primarily red, brown, yellow, white to gray, often with colorless layers

Occurrence: Uncommon

Notes: At around a billion years old, Lake Superior's agates are among the world's oldest, and virtually every specimen will show some sign of weathering. Some fared better than others, of course, and among the most unfortunate examples are known as peeled agates. "Peelers," as collectors refer to them, are agates that have separated or broken along, rather than across, their banding. As solid and tightly bonded as an agate's bands may seem, they actually contain microscopic pores and spaces between them. Once water has made its way into those pores, it can freeze and thaw again and again, slowly weakening the bonds between bands. Eventually—often with the help of a glacier or a river—the weak bond can fail, causing upper layers to "peel" off of lower ones. This leaves very smooth surfaces behind; just like peeling back the outer layers of an onion, these surfaces are actually the surfaces of the interior agate bands. This is one of the most dramatic kinds of weathering in agates and can produce strange and interesting shapes.

Where to Look: Since peeled agates are derived from weathering, it's hard to predict where they'll turn up. Many specimens with significant peels are found in Duluth, Minnesota, area gravel pits and even fields and ditches. Elsewhere, rivers tend to be places where weathering knocks the layers of agates apart.

Rough sagenitic agates

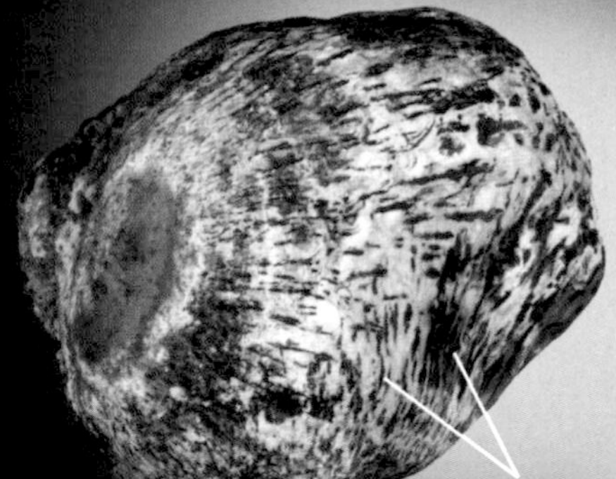

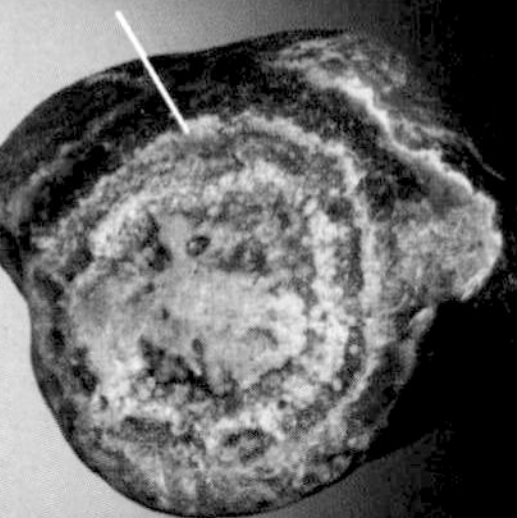

Sagenite formations chalcedony

Sagenite formation with calcite center

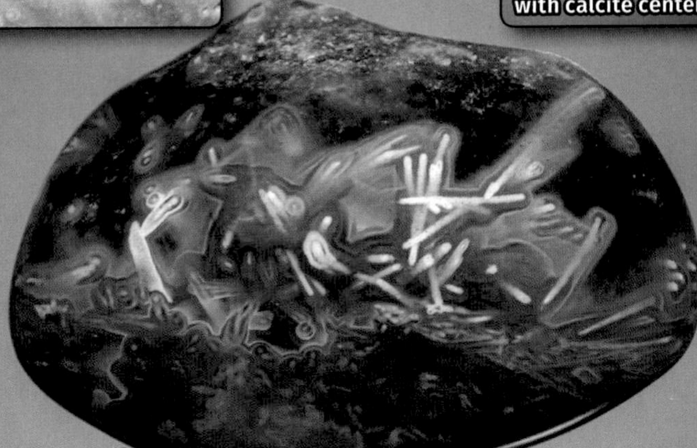

Polished sagenitic agate

Agates, sagenitic

Hardness: 6.5–7 **Streak:** White

Primary Occurrence

Environment: Shores, rivers, pits

What to Look for: Very hard, waxy, red-brown translucent masses containing radiating groups of needle-like inclusions

Size: Sagenite growths can be seen in agates of any size

Color: Multicolored; varies greatly, but primarily red, brown to yellow, white to gray; inclusions may also be black

Occurrence: Uncommon

Notes: Lake Superior agates often contain inclusions, or growths of other minerals that competed for space within the vesicle (gas bubble) when an agate was forming. Most are subtle or mundane, but occasionally they can become the main feature. Such is the case in sagenitic agates (called "sagenites" by collectors), which contain clusters of needle-like mineral inclusions, sometimes criss-crossed in seemingly random patterns, but more often arranged into radial, fan-shaped or circular patterns extending into the chalcedony. The minerals responsible for sagenitic growths are usually zeolites (page 223) and goethite (page 155)—many delicate "sprays" of crystals are assumed to be a zeolite, while coarser ones are assumed to be goethite (though this is not always correct). In most cases, however, the original mineral is largely gone, replaced in part or fully by quartz. Due to their heavily included nature, sagenitic agates often contain small voids or pores, making them appear a bit rough, and most will greatly benefit from polishing. However, more often than not, typical agate banding is subtle or absent altogether.

Where to Look: Sagenites can be found on many of the western half of Lake Superior's shores, but the greatest concentration is in Minnesota. Gravel pits and rivers in the Duluth area (and further southwest) have produced fantastic specimens.

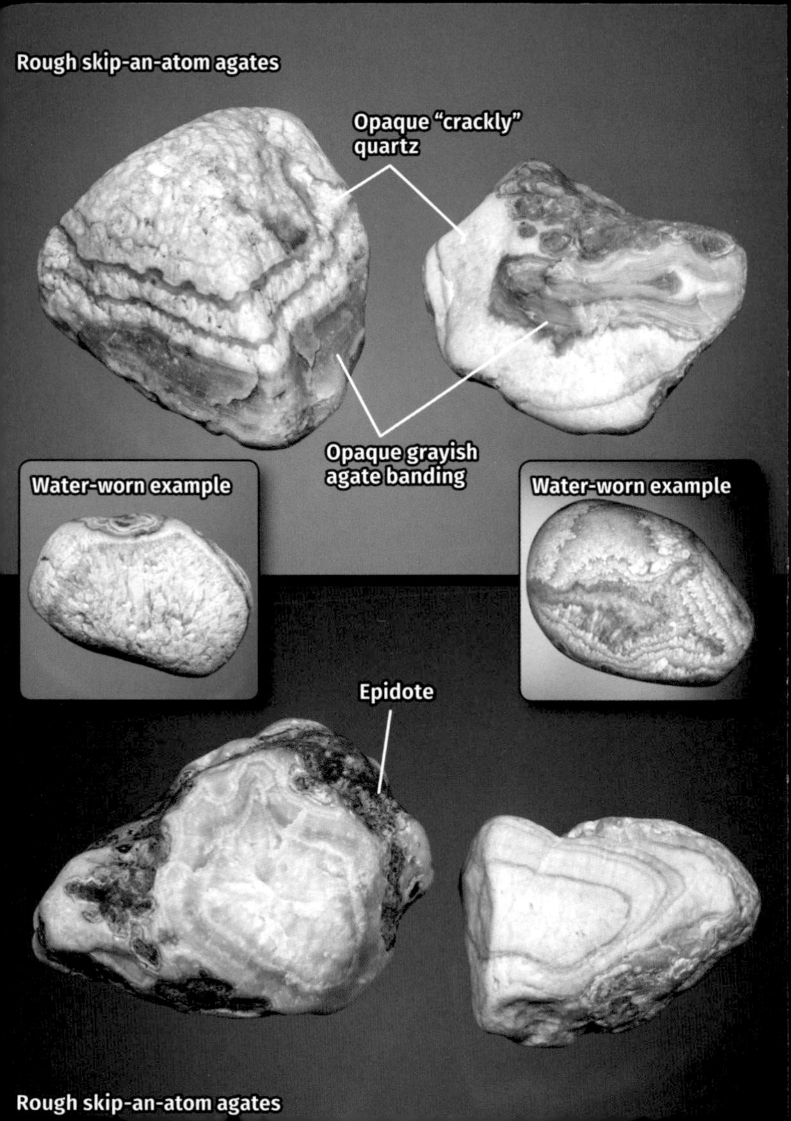

Rough skip-an-atom agates

Opaque "crackly" quartz

Opaque grayish agate banding

Water-worn example

Water-worn example

Epidote

Rough skip-an-atom agates

Agates, skip-an-atom

Hardness: 6.5–7 **Streak:** White

Primary Occurrence

Environment: Shores, rivers

What to Look for: Very hard masses containing pale agate banding and opaque white or bluish gray quartz

Size: Most specimens are no larger than a softball

Color: Multicolored; varies, but banding is often white to tan or brown; quartz is white to gray, sometimes bluish

Occurrence: Rare

Notes: Lake Superior is host to more variation in its agates than virtually any other place in the world, but perhaps none are as odd and unique as "skip-an-atom" agates, known among collectors as "skips." These agates contain large amounts of opaque white or bluish gray quartz, often with a "crackled" or fragmented appearance and arranged into layered growths. The quartz is often interspersed with agate banding, though usually sparing and of distinctly pale, nearly opaque colors, particularly brown or grayish (very rarely reddish). Most are found weathered free of their host rock, but others can be found still tightly embedded nodules in basalt and, to a lesser extent, as fillings in irregular cavities in porphyry (page 185). How they formed is a mystery—a leading idea is that these agates were altered by heat after they had already formed—and their strange appearance has led to equally strange theories: the tongue-in-cheek suggestion that the quartz "skipped an atom" during formation gives them their name, but it makes little scientific sense.

Where to Look: These odd agates can rarely be found on many Minnesota, Wisconsin, and Michigan beaches, but they aren't abundant anywhere aside from Minnesota beaches between Duluth and Silver Bay. They have also been found to a lesser extent in the Whitefish Bay, Michigan, area.

Calcite-rich material
Rough Thunder Bay agates
Desirable orange color
Lacy patterning
Sawn example
Attached host rock
Rough Thunder Bay agates

Agates, Thunder Bay

Hardness: 6.5–7 **Streak:** White

Primary Occurrence

Environment: Mines, outcrops, pits

What to Look for: Hard, irregular masses of quartz and agate banding with thin, lace-like patterns in limestone

Size: Thunder Bay agates can range from pebbles to boulders

Color: Multicolored; varies, but usually brown to tan, yellow to orange or reddish, white to gray or black

Occurrence: Rare in Ontario; very rare to nonexistent elsewhere

Notes: One of the scarcest and most unique varieties of agates in the Lake Superior region are the vein agates (read more on page 81) found in a small localized area near Thunder Bay, Ontario. These mostly opaque, brown agates did not form in vesicles (gas bubbles) within volcanic rocks, but rather in blocky cavities and cracks developed within a body of soft, iron-rich limestone. As a result, they contain many mineral inclusions, particularly calcite and iron-bearing minerals as well as quartz stalactites (icicle-like formations), all of which caused the chalcedony banding to conform to them, producing lace-like patterns. True agate banding is usually sparing, outnumbered by coarse quartz crystals, but it can be attractive and complex. Many specimens will have some attached limestone, often indurated (hardened) by quartz, and they can vary greatly in size, with some rare specimens being massive boulders several feet across.

Where to Look: With very rare exceptions of water-worn pieces transported to northern Minnesota, all Thunder Bay agates are found in their namesake region. Unfortunately, the chief area for these agates, just a few miles northeast of the city of Thunder Bay, is largely privately owned and the pay-to-dig mine once operating there is now closed. Nearby rivers and gravel pits do produce smaller specimens, however.

Minnesota thunder eggs

Whole, unbroken example

Irregular layered chalcedony

Clay-lined interior (white)

Cut example

Quartz

Pockets of agate banding

Cut and polished thunder egg

Specimen courtesy of Dave Woerheide

Agates, thunder egg

Hardness: 6.5–7 **Streak:** White

Primary Occurrence

Environment: Rivers, outcrops

What to Look for: Rough, brown, approximately spherical rocks containing quartz and reddish chalcedony within

Size: Most are fist-size, though some may be larger

Color: Brown exterior; multicolored interior, including red to orange, pinkish to white, and tan to brown

Occurrence: Rare: in the Lake Superior region, only found in Minnesota

Notes: Thunder eggs are a popular variety of agate renowned in the American West, but Minnesota's shores of Lake Superior are also home to at least one occurrence of this important type of agate. Thunder eggs formed within bodies of tuff, a volcanic rock formed of fused volcanic ash. There is no tuff in the Lake Superior region today—it is a fairly soft rock that degrades into clay when it weathers, and the glaciers easily wiped away all trace of it—but the existence of thunder eggs indicates that it was once present. Minnesota's thunder eggs are rough, ball-shaped masses of rock that resemble rhyolite or hardened clay on their exteriors but contain jagged pockets of quartz and agate banding within. Some are hollow geodes, lined with quartz crystals or clay, while others are filled completely. The agate banding may be poorly developed, but is almost always iron-rich shades of orange and red. They are a fairly rare find today; most are found along rivers where they blend in with the muddy banks. Identification is always tricky until cut or broken open, since they look like common rock when whole.

Where to Look: These rarities are found just a few miles northeast of the Grand Marais, Minnesota, area. They aren't typically found on the shore; try the muddy banks of adjoining rivers instead.

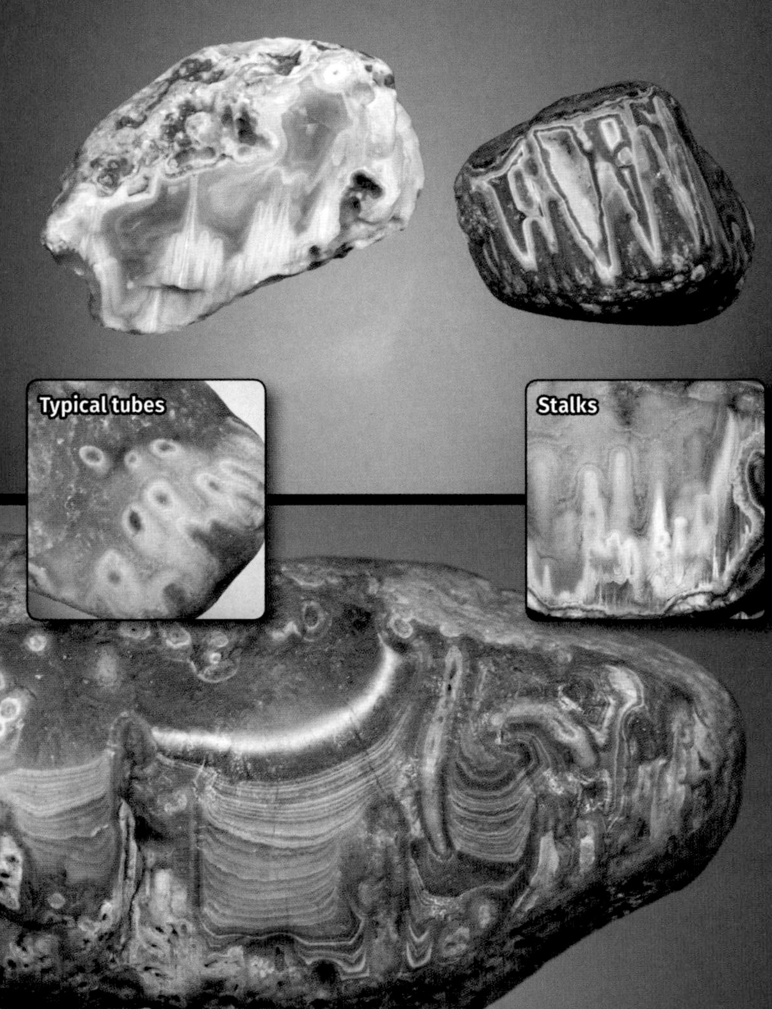
Rough tube agates
Typical tubes
Stalks
Rough agate with curving tubes

Agates, tube & stalk

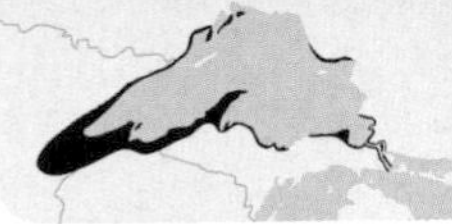

Primary Occurrence

Hardness: 6.5–7 **Streak:** White

Environment: Shores, rivers, pits, outcrops

What to Look for: Very hard, waxy masses containing agate banding and long, generally cylindrical structures

Size: Specimens can range from pebbles to fist-size or larger finds

Color: Multicolored; varies greatly, but banding is primarily red, brown, yellow, white to gray, often with colorless layers

Occurrence: Uncommon

Notes: A large percentage of Lake Superior agates contain some kind of mineral inclusion, and tubes and stalks are among the most easily identifiable. Tube agates contain cylindrical structures that extend through the chalcedony, often forcing the agate banding to conform to their shape. These generally formed when chalcedony coated long, slender crystals of other minerals and embedded them within the nascent agate. As their name implies, these can be hollow tubes, but they are more often filled in with quartz. You may find small dimples or "pin holes" where tubes meet the surface of an agate. Occasionally tubes can be beautifully banded and look similar to an eye agate (page 47), and may have a matching "eye" on the opposite side of the specimen (if the tube extends all the way through). Stalks are similar, but tend to be tapered in shape and only extend partway through an agate. These formed when low-density mineral inclusions slowly floated upward through the soft body of a developing agate. In both cases, the inclusions are usually straight and parallel, but they are more rarely curved or "bent," possibly caused by movement of the host rock during agate formation or changes in pressure within the developing agate.

Where to Look: Tubes and stalks are typical inclusions found in many Lake Superior agates and can occur all over the region. A particularly lucrative area is the Duluth, Minnesota, area gravel pits and river beds all along Minnesota's shore.

Paint agate vein in host rock

Chlorite-filled vesicles

Dark vein agate

Agate vein in jasper

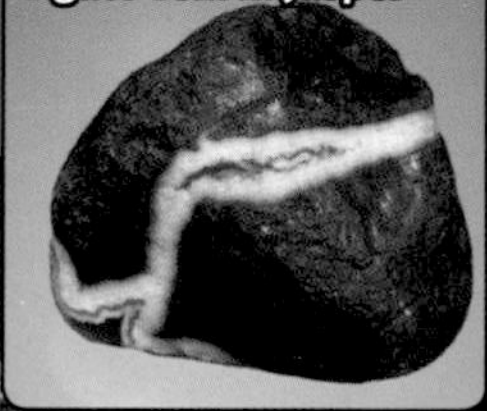

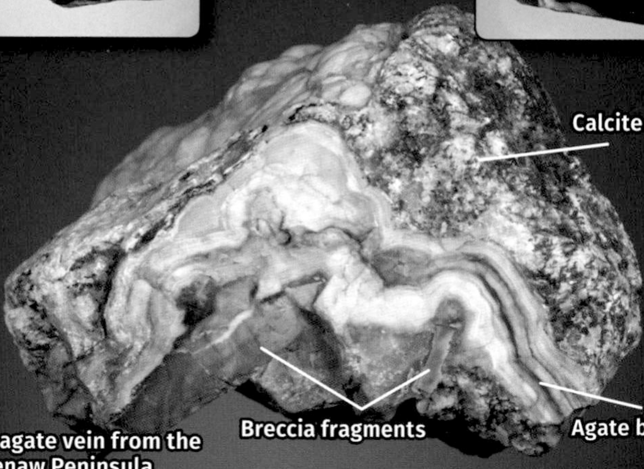

Calcite coating

Breccia fragments

Agate banding

Large agate vein from the Keweenaw Peninsula

Agates, vein

Hardness: 6.5–7 **Streak:** White

Primary Occurrence

Environment: All environments

What to Look for: Very hard, waxy, elongated veins or masses containing agate banding

Size: Vein agates aren't usually much longer than 3 to 5 inches

Color: Multicolored; varies greatly, but banding is primarily red, brown, yellow to orange, white to gray or black

Occurrence: Rare

Notes: While the majority of Lake Superior's agates formed inside vesicles (gas bubbles) within volcanic rocks, others were able to develop within cracks and other irregular cavities in a variety of rocks. These are called vein agates, and they exhibit elongated, thin banded patterns derived from forming in this setting. Many are found in typical agate-bearing volcanic rocks like basalt and rhyolite, but others can be found in chert (or jasper) and, in the case of Thunder Bay agates (page 75), in limestone. In most examples, you will not see concentric banding, but rather more-or-less parallel banding along the length of the vein (though occasional isolated spots of ring-like banding do occur). Depending on how a specimen has weathered, or if it has been freed from its host rock, the vein-like nature of the agate banding may not be obvious; rounded, water-worn examples have lost much of their shape and context. Very elongated, "stretched" vesicles may produce vein-like agates, but they typically have smoother, more regular edges, in contrast with the often jagged shape of vein agates' host cavities.

Where to Look: Some rare agate veins have been found around the tip of the Keweenaw Peninsula, even including some examples found in submerged rock. Black vein agates are found in Minnesota gravel pits and rivers all along the shore.

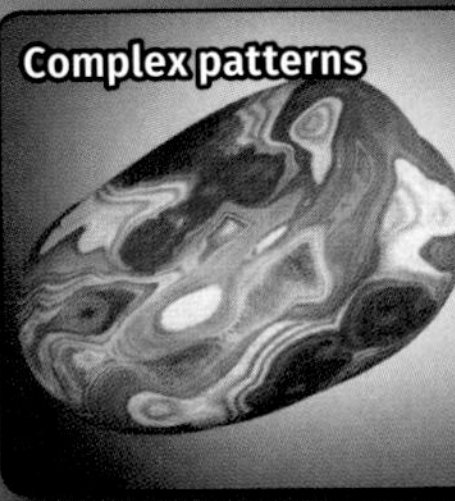

Water-washed agates from eastern Michigan

Specimens courtesy of Mark Bowan

Primary Occurrence

Agates, water-washed

Hardness: 6.5–7 **Streak:** White

Environment: Shores, rivers

What to Look for: Very hard, waxy masses containing agate banding and smooth, rounded surfaces

Size: Water-washed agates are usually fist-size or smaller

Color: Multicolored; varies greatly, but banding is primarily red, brown, yellow, white to gray, often with colorless layers

Occurrence: Uncommon

Notes: Minerals are rarely enhanced by millennia of weathering, but agates' high hardness means that they are able to gracefully endure far more abuse from waves and ice than most other gemstones. Water-worn agates, known locally as "water-washed" agates, are beautiful examples of how the harsh elements can wear down the thick outer layers of an agate to reveal the often wild and complex banding below. In many of these sought-after agates, weathering has not just revealed one banded face of the agate, but rather has exposed the banded pattern on all sides of a specimen. As their name suggests, they've spent countless years in rivers and lakes, becoming rounder and smoother in shape as sand and other rocks roll against them—though it was the glaciers of the past Ice Ages that began this sculpting process. As such, water-washed agates have no sharp corners or hard edges, and some have become so finely worn that they can appear exceptionally glossy, almost as if naturally polished. Brightly colored specimens with bold patterns that wrap around all sides of the agate are among the most valuable agates.

Where to Look: Water-washed agates can be found almost anywhere; Minnesota's shores, especially from Two Harbors to Silver Bay, are lucrative, as are Wisconsin's shores around Saxon Harbor. Ontario's shores, around Rossport to Terrace Bay and nearby islands, have produced smoothed agates, as have Michigan beaches east of Grand Marais.

Rough whorl agates

Wispy, unevenly colored banding

Whorl banding

Gravitational layers

Polished whorl agate

Odd, non-parallel structures

Polished "cloud" agates

Agates, whorl

Hardness: 6.5–7 **Streak:** White

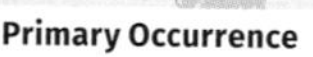

Primary Occurrence

Environment: Shores, rivers, pits

What to Look for: Very hard, waxy masses containing agate-like banding consisting of broad, sweeping shapes

Size: Most whorl agates are smaller than your fist

Color: Multicolored; varies, but typically brown to orange, sometimes bright red; also white to gray

Occurrence: Uncommon to rare

Notes: As if attempting to study and explain agate formation weren't already complex and difficult enough, varieties like whorl agates only further confound the issue. Also known as "cloud agates" or "veil agates," whorl agates contain very strange banding that appears lumpy or wavy, ropy, and often unevenly spaced, with sweeping, "billowing" shapes. In some, this odd banding clearly formed first and is relegated to the edges of the specimen while containing another more typical type of banding or quartz in the center. In others, it dominates the agate with thick, sometimes overlapping shapes reminiscent of water-level agates (page 57). Whatever their cause, it could be that their formation was not a consistent process and may have been affected by outside chemicals or forces. In most whorl agates, coloration is often on a gradient, where the iron-derived reds and browns grow and fade in intensity along the bands, and in others, dark grays fade to white. These odd agates have not been studied in detail and occur in very few places around the world, making Lake Superior's whorl agates a fortunate yet curious find for area collectors.

Where to Look: Whorl agates are not abundant in any one area, though the gravel pits in the Duluth, Minnesota, area (and further southwest) produce examples fairly regularly.

Domeykite veins (dark) in metasedimentary rock
"Mohawkite" (dark) in stained quartz
"Mohawkite" veins in cut quartz
Metallic mass of "mohawkite"
Bluish iridescence on "mohawkite"
Brassy "mohawkite" veins in quartz

Algodonite & Domeykite

Primary Occurrence

Hardness: 3–4 **Streak:** Dark gray

Environment: Mines, outcrops

What to Look for: Metallic veins, yellowish or gray in color, within quartz or quartz-rich rock near copper mines

Size: Veins of these minerals can measure several inches to a foot

Color: Silvery gray to pale brassy yellow; bluish or greenish surface coatings when oxidized; sometimes dark gray

Occurrence: Rare in Michigan; very rare in Ontario; not found in Wisconsin or Minnesota

Notes: Algodonite and domeykite are two separate minerals with such a similar composition and means of formation that they are virtually impossible to tell apart outside a lab. They also usually occur together, inseparably intergrown in the same masses. Crystals of either mineral are extremely rare and instead generally only appear as veins or masses, usually embedded within quartz. Both minerals are composed of a combination of copper and arsenic; in Michigan, the blend of these two minerals, often with some admixed native copper, are known as "mohawkite," the local name given to this once-mined copper ore. When freshly exposed, both minerals will look like a silvery gray metal, but on exposure to air they will change color, first to a brassy yellow color, often tinged with greenish iridescent surfaces, then to a dark, dull gray. While they can appear similar to pyrite, algodonite and domeykite are much softer; chalcocite and silver may also appear similar, but both are softer. Though they contain arsenic, algodonite and domeykite are largely safe to collect. However, the dust they produce can be harmful if ingested, so gloves and a mask are recommended.

Where to Look: Keweenaw Peninsula mines in Houghton and Keweenaw Counties, particularly around Mohawk-Ahmeek, are the primary locality. The Pancake Bay and Batchawana Bay, Ontario, area metasedimentary rocks may also rarely contain dark veins of domeykite.

Minerals within a cavity in metamorphosed basalt

Yellow andradite garnets

Actinolite or tremolite fibers

Hornblende crystal in host rock

Flat, perfectly cleaved surface

Hornblende in syenite

Large, embedded, glassy hornblende crystal

Amphibole group

Primary Occurrence

Hardness: 5–6 **Streak:** White

Environment: All environments

What to Look for: Usually dark-colored, fibrous or glassy crystals, masses, or grains embedded in rocks like granite

Size: Most amphibole crystals are thumbnail-size or smaller

Color: Dark yellow to green, brown, gray to black, reddish

Occurrence: Common

Notes: Like the pyroxene group (page 195), the amphiboles are a family of rock-builders, meaning that they're most prevalent as constituent minerals in rocks as embedded grains or crystals. In particular, they're most common in Lake Superior's igneous rocks, like granite, and some metamorphic rocks. Several amphiboles are present in the region, including tremolite, actinolite, and especially hornblende (a general term for any dark amphibole), and most examples will show a silky, fibrous luster or texture along their length. (Though very finely crystallized hornblendes may appear smoother and more brightly lustrous). Because they are often a tightly embedded component of rocks, most are easy to overlook unless clear crystal shapes are present. Black hornblende crystals are generally rectangular with angular ends, conspicuous within granitoids; lighter colored tremolite and actinolite are scarcer, found as bundles of fibers in metamorphosed rocks. Distinguishing amphiboles from pyroxenes is tricky; amphiboles' generally fibrous luster and nearly perfect cleavage (they break along a perfectly flat plane) are helpful. Though fairly rare in the region, some amphibole group minerals are asbestiform: they crystallize as tiny, fiber-like crystals called asbestos. These varieties are rare, but if you suspect them, wear a respirator, as inhaling the tiny fibers is dangerous.

Where to Look: The coarse-grained Canadian Shield rocks found along Ontario's shores have opportunities to find embedded crystals. Syenite pebbles from eastern Michigan can also have nice embedded crystals.

ough anorthosite
Reflective embedded feldspar crystals
Close-up of texture
Darkly colored specimen
ough specimen of richly colored anorthosite

Anorthosite

Hardness: ~6–6.5 **Streak:** N/A

Primary Occurrence

Environment: Shores, outcrops

What to Look for: Light-colored, greenish rock composed predominantly of coarse, glassy crystals

Size: Anorthosite can occur in any size, from pebbles to cliffs

Color: Light gray to gray-green or dark green, rarely pink

Occurrence: Uncommon in Minnesota; very rare elsewhere

Notes: Anorthosite is a unique coarse-grained igneous rock in that it consists almost entirely of one mineral: plagioclase feldspar (page 139). Less than 10 percent of the rock consists of other minerals. It is also not entirely understood how it forms; we know it originated very deep within the earth, but the process by which it developed almost exclusively from one mineral is still debated. Due to the extreme depth of its formation, outcroppings of anorthosite are rare around the world and pose interesting questions for researchers, such as which geological processes forced blocks upward from such a depth. The wider Lake Superior region is fortunate to have several important outcroppings, but the only ones closely associated with the lakeshore are in Minnesota. Comprised of coarse, glassy and translucent interlocked crystals, anorthosite may initially resemble gabbro (page 149), but it lacks most of the opaque, dark-colored grains present in gabbro, in addition to anorthosite being far rarer. Most anorthosite is pale to dark green, or gray-green, but it tends to turn whiter and more opaque as it weathers.

Where to Look: Minnesota's Lake Superior shores is the primary place in the region to find anorthosite. The Silver Bay area is home to several outcrops along Highway 61, and the famous Split Rock Lighthouse sits on a massive block of anorthosite. Other places may yield glacially deposited pieces.

“Jaspilite” specimens, containing jasper (red) and hematite (gray)

Upper specimens courtesy of John Woerheide

Right inset specimen courtesy of Dave Woerheide

Common appearance

Water-worn sample

Water-worn example

Subtle, parallel layers

Rough example

Specimens courtesy of Phil Burgess

Banded Iron Formation

Hardness: 6–7 **Streak:** N/A

Primary Occurrence

Environment: Outcrops, pits, rivers, mines

What to Look for: Very hard, layered masses with differently colored opaque, dense layers, particularly red and gray

Size: Masses can range from inches to feet in size

Color: Multicolored; varies, but usually with red to brown layers and gray to black layers that often show a metallic luster

Occurrence: Uncommon

Notes: The Lake Superior region's massive iron deposits are largely present as Banded Iron Formations, or BIFs, for short. BIFs are rocks that formed billions of years ago at the bottom of ancient seas when there was far less oxygen in the atmosphere than today. Cyanobacteria, such as stromatolites (page 147), were among the first organisms on Earth, and the microscopic organisms played a key role in the early oceans, converting the harsh atmosphere to a more hospitable, oxygen-rich one through photosynthesis. As oxygen was produced, it combined with dissolved iron and silica in the water, forming particles that then sank into thick, muddy beds. Over millions of years, these beds compressed to form tightly layered, iron-rich BIFs. The most common appearance of BIFs in the region is a red jasper (page 165) with parallel layers of dark, sometimes "glittery" hematite throughout, known locally as "jaspilite." Jaspilite is one of the most collectible and easily identified BIF rocks; it was transported by glaciers and is often found along rivers and in gravel pits. Taconite (page 217) is also an important BIF, but it is less collectible.

Where to Look: Minnesota's shores, particularly in the Grand Marais area, are known for specimens. Chunks will also turn up in northern Wisconsin, around the Gogebic Range on the Michigan border, and in Michigan, around Ishpeming.

Large crystal cluster from a Minnesota lakeside rhyolite cliff
Baryte blades
Calcite
Translucent baryte from Michigan mine
Detail of crystal surface
Small baryte crystals (tan) on amethyst (purple) from Ontario

Baryte (Barite)

Hardness: 3–3.5 **Streak:** White

Primary Occurrence

Environment: Outcrops, pits, mines

What to Look for: Light-colored, thin, brittle, blade-like crystals, often intergrown in complex clusters

Size: Individual crystals are rarely larger than an inch or two; masses or crusts may be several inches in size

Color: Colorless to white or gray; often tan to brown or reddish

Occurrence: Uncommon to rare

Notes: Baryte is the world's most abundant barium-bearing mineral. It is rare in the Lake Superior region, but the localities that do produce it can yield fantastic crystallized specimens. Baryte can form in a wide range of geological environments: it is often associated with sedimentary rocks where it generally occurs as uninteresting granular masses, but it also develops within metallic ore deposits (such as those in iron and manganese mines) as fine crystals and within cavities in volcanic rocks. Its crystals can also show considerable variation in shape, but in the Lake Superior region, tabular (flat, plate-like) bladed crystals are most common. These crystals are usually rectangular with flat tips and beveled edges, often showing many smaller step-like growths on their surfaces. They're sometimes translucent but more often opaque, and generally always light-colored, though frequently stained brownish by iron. Crystals are also very brittle and soft, but all share baryte's distinctive high density; even small crystals will feel heavy for their size.

Where to Look: Ontario's amethyst mines produce fine, opaque baryte atop quartz crystals. In Michigan, the Copper Harbor Conglomerate produces baryte in Ontonagon County up through the Keweenaw Peninsula. In Minnesota, rhyolite cliffs on the shore near Duluth produce crystal clusters.

each-worn basalt
Mineral vein
Vesicles
Mineral-lined
vesicles
Weathered basalt
with zeolites (orange)
Reddish basalt with
countless vesicles
Even coloration
Calcite in
elongated vesicles
Freshly broken
surface
each-worn basalt

Basalt

Primary Occurrence

Hardness: 5–6 **Streak:** N/A

Environment: All environments

What to Look for: Abundant dark-colored rock with a fine-grained texture, often with many vesicles (gas bubbles)

Size: Basalt can be found in any size, from pebbles to cliffs

Color: Gray to black, often with a greenish or red-purple tint

Occurrence: Very common

Notes: Basalt is among the most common volcanic rocks around the world and forms when lava (molten rock) rich with iron and magnesium erupts onto the earth's surface and cools rapidly. When the Midcontinent Rift began to split North America apart across what is now the Lake Superior region, enormous amounts of basalt lava rose from the earth to fill the void through successive eruptions. The rapid cooling of basalt lava means that the minerals within it—largely plagioclase feldspars, olivine, pyroxene minerals, and magnetite—had very little time to crystallize, remaining as small grains. If the same body of molten rock had remained inside the earth where it could cool slowly, it would've become gabbro (page 149). Today, massive basalt flows can still be seen throughout the region, appearing as very dark, fine-grained rock with an even-colored appearance. The rock's rapid cooling also means that lots of gas was trapped in it as vesicles (gas bubbles) in which minerals, such as agates and zeolites, later formed amygdules (rounded mineral formations). It can resemble rhyolite (page 205) but is usually darker, more finely grained, and can be weakly magnetic.

Where to Look: Glacially deposited pebbles are found amply on nearly any shore on the western half of the lake (to a lesser extent on the eastern half), and huge flows can still be seen on the shores in Minnesota and along the Keweenaw.

Pearly sheen
"Dog tooth" crystal
Complex
rhombohedral crystal
alcite rhombohedrons on host rock

Calcite, crystallized

Hardness: 3 **Streak:** White

Primary Occurrence

Environment: All environments

What to Look for: Abundant, soft, light-colored blocky crystals, masses, or veins that are easily scratched with a US nickel

Size: Crystals are usually thumbnail-size, but may rarely be fist-size or larger; masses or veins can grow to several feet

Color: Colorless to white, gray; often stained yellow to brown

Occurrence: Very common

Notes: Calcite is one of the most common minerals on Earth and forms in most geologic environments, making it prevalent all around Lake Superior. This also means that it is one of the most important minerals to learn to identify when rock hounding in the region. Worldwide, calcite can develop in a startling number of crystal shapes—around 800—but just a few are prevalent in the Lake Superior region. Most common are rhombohedral crystals, which are shaped like "leaning" cubes, but more complex, modified rhombohedral shapes are common as well and may appear like sharply angular balls or like rhombohedrons that have had their corners cut off. Steeply pointed, six-sided "dog tooth" crystals are also common. But as with any common mineral, most abundant of all are irregular masses or veins that filled a cavity completely. Water-worn pebbles turn up on beaches and in rivers as well. But whatever form it takes, calcite is always easy to identify due to its distinctive crystal shapes, low hardness, and pearly, "flashy" sheen on freshly broken surfaces.

Where to Look: Calcite is ubiquitous in the Lake Superior region and you'll find it virtually anywhere there is exposed rock. Fine crystals are frequent finds in volcanic rocks along rivers and cliffs in Minnesota and the Thunder Bay, Ontario, area. Mine dumps in Michigan often produce small, sharp crystals and coarse veins, as do northeastern Wisconsin gravel pits.

Freshly broken calcite vein
Water-worn granular calcite
Cleaved, broken rhombohedrons
Calcite-filled steam channels in basalt
Crystal fluorescing in short-wave UV light
Fluorite (purple)
Calcite vein in rhyolite

Calcite, masses & veins

Hardness: 3 **Streak:** White

Primary Occurrence

Environment: All environments

What to Look for: Soft, light-colored blocky crystals, masses, or veins within cavities in volcanic or sedimentary rocks

Size: Crystals are usually thumbnail-size, but may rarely be fist-size or larger; masses or veins can grow to several feet

Color: Colorless to white, gray; often stained yellow to brown

Occurrence: Very common

Notes: Whether finely crystallized or in a less ideal form, calcite is not difficult to identify, thanks to its easily testable traits. Calcite is typically colorless to white but is frequently tinted yellowish to brown by iron or clay. Most specimens will exhibit a glassy luster when well-developed or freshly broken, but more weathered samples will be dull. Some masses that formed more granularly rather than as crystallized masses will be dull and opaque, quite resembling datolite (page 133) but are softer. In fact, calcite's low hardness is a key characteristic no matter what its form. Another important trait to learn to spot is its perfect rhombohedral cleavage. This means that when struck or broken, any specimens of calcite will break at specific angles resembling an angled or "leaning" cube; highly weathered examples often show many planes of cleavage, giving their surfaces the appearance of steps. Veins and masses will superficially resemble quartz (page 197), but quartz is far harder. Calcite will also effervesce, or fizz, in even weak acids like vinegar, which will distinguish it from baryte (page 95).

Where to Look: The various basalt and rhyolite lava flows of the region, from Thunder Bay, Ontario, down to Duluth, Minnesota, and across northern Wisconsin and the Keweenaw Peninsula of Michigan, will exhibit countless pockets and veins of calcite, often with other minerals alongside. Carefully splitting the stone at its natural cracks and fissures and checking muddy clay-filled pockets may yield fine crystals.

Common beach-worn examples

Conchoidal (curved) broken surface

Pitted, unbroken surfaces

Waxy luster

Chalcedony nodules in basalt

Common beach-worn examples

Chalcedony

Hardness: 6.5–7 **Streak:** White

Primary Occurrence

Environment: All environments

What to Look for: Very hard, waxy, translucent masses of material often with mottled brown to red coloration

Size: Masses of chalcedony can vary, rarely up to boulder-size

Color: Brown to red, white to yellow, gray; often multicolored

Occurrence: Common

Notes: Chalcedony is a common variety of microcrystalline quartz, which means that although it is identical to quartz (page 197) in most ways, its crystals are too small to see without a powerful microscope. Chalcedony's microcrystals differ from those found in jasper (page 165) and chert (page 109) in that they are shaped like tiny parallel fibers, which makes it translucent when thin—a key difference between it and opaque jasper and chert. Like all microcrystalline quartz, it does not form outward shapes of its own, but rather takes the shape of its surroundings during formation. This may be in cracks or veins within rock, or as frequently occurs in the Lake Superior region, within vesicles (gas bubbles) in volcanic rocks. Chalcedony is best known for being the material from which agates (page 39) and their trademark concentric bands are formed, but common chalcedony is usually not as interesting. Most is colored in shades of brown and red, derived from iron impurities, but "purer" specimens may be white or gray, often with some mottling or variegation. Like all forms of quartz, chalcedony is very hard, has waxy surfaces, a conchoidal fracture (curved breaks), and has very sharp edges when freshly broken.

Where to Look: Virtually any beach or river will yield countless little pebbles of reddish chalcedony. Larger masses turn up in any gravel pit, especially in the Duluth, Minnesota, area.

Large chalcocite vein in host rock
Malachite (green)
Chalcocite (gray)
Chalcocite in host rock
Metallic chalcocite mass

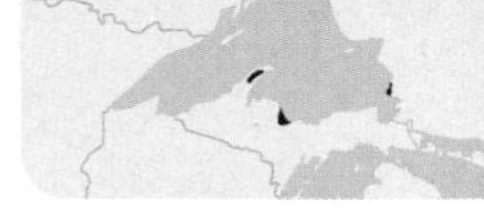

Primary Occurrence

Chalcocite

Hardness: 2.5–3 **Streak:** Metallic gray-black

Environment: Mines, outcrops

What to Look for: Dark, metallic gray mineral found as veins or masses in rock, often with associated blue or green minerals

Size: Specimens may measure several inches to feet in size

Color: Metallic steel-gray to black, bluish black

Occurrence: Rare in Michigan and Ontario; very rare in Wisconsin; not found in Minnesota

Notes: Chalcocite is a simple combination of copper and sulfur, and as such, it has long been mined as an ore of copper. It is fairly widespread but in small amounts and is only abundant at certain Michigan mine sites. It can form a number of ways, including as a direct result of hydrothermal (hot, mineral-bearing groundwater) activity that deposits minerals into cavities in rocks. But in this region, it occurs more frequently as a secondary mineral that developed after copper and copper-bearing minerals underwent weathering and chemical change, which freed copper atoms from them that could then contribute to new minerals. This makes chalcocite primarily a mineral found in veins, as it formed as masses within cracks in various kinds of rocks, primarily igneous and sedimentary types. Fine crystals are extremely rare. Identification won't be difficult, as algodonite and domeykite (page 87) are the only minerals that appear similar, but both are slightly harder than chalcocite. In addition, chalcocite often has some blue chrysocolla or green malachite associated with it.

Where to Look: Chalcocite is not a typical find along Lake Superior. Virtually all specimens will come from igneous rocks in Keweenaw Peninsula mine dumps, veins in some sedimentary rocks near Marquette, Michigan, and some old mines and outcrops in the Mamainse Point, Ontario, area.

Chalcopyrite (brassy yellow) in quartz
Chalcopyrite with rusty stain in quartz
Right inset specimen courtesy of Eric Powers
Chalcopyrite in quartz

Chalcopyrite

Primary Occurrence

Hardness: 3.5–4 **Streak:** Greenish black

Environment: Outcrops, pits, mines

What to Look for: Brittle, soft yellow metallic masses or veins in rock or quartz, often with a brown or greenish surface stain

Size: Masses are rarely larger than an inch or two

Color: Brassy yellow to golden yellow, metallic brown; may have rusty brown or iridescent surfaces when weathered

Occurrence: Uncommon

Notes: Chalcopyrite is a metallic mineral very similar to pyrite (page 193), but it also contains copper along with its iron content. Crystals are very rare in the typical settings in which you'll find it around Lake Superior; instead, it often appears as embedded masses or grains in a variety of rocks. It can be found in some limestone formations around the lake, but it may be more frequently encountered in igneous rocks, like gabbro (page 149) where it appears in irregular pockets, and in quartz (page 197) as veins or small masses. In the majority of cases, growths of chalcopyrite won't show any particular shape, instead filling the cavities available to it. Its bright metallic yellow coloration makes it particularly conspicuous, and even when poorly formed, you're not likely to miss it. All of these traits will make it easily confused with pyrite, but pyrite is generally paler in color and is always harder. In addition, pyrite may be found crystallized more frequently and takes on a cubic shape not shared by chalcopyrite. Lastly, when weathered, chalcopyrite can develop colorful iridescent surfaces, and when it degrades it produces rusty yellow stains and growths of green malachite (page 175).

Where to Look: Chalcopyrite can be found, albeit rarely, all over the region; it turns up in Keweenaw mine dumps, Minnesota gabbro, and veins in Thunder Bay, Ontario, region granitoids (granite and granite-like rocks).

Beach-worn chert
Chert layers (black) with siltstone (brown)
Example colored by limonite
Stromatolite-bearing samples
Rough sample
Coral fossil in chert
Layered chert
Iron-rich chert
Beach-worn chert

Primary Occurrence

Chert

Hardness: 7 **Streak:** White

Environment: Shores, rivers, pits, outcrops

What to Look for: Very hard, opaque masses, often with smooth, waxy surfaces and sharp edges when broken

Size: Chert can be found in any size, from pebbles to boulders

Color: Varies; gray to black, white, tan to yellow, brown to red

Occurrence: Very common

Notes: Chert is a sedimentary rock that formed on ancient seafloors when thick beds of silica-rich mud were compacted and solidified. As such, it consists almost entirely of quartz in the form of microscopic grains, along with minor amounts of other minerals and color-causing impurities. The result is a dense rock that is so hard that typical weathering from wind and waves effectively polishes it, bringing out its wax-like luster. Like other quartz-based material, it also exhibits conchoidal fracture (when struck, circular cracks will appear) and breaks with very sharp edges. It is usually gray to brown in color and may exhibit colored layers, or it can be found with alternating layers of other rocks entirely, such as siltstone or iron ores, as in banded iron formation (page 93). It can also contain fossils, particularly coral or algae (see stromatolites, discussed separately on page 147). It can be confused with chalcedony (page 103) and agates (page 39), but those are translucent and agates have concentric bull's-eye rings. Black chert is called flint, and colorful varieties are called jasper (page 165).

Where to Look: Chert is ubiquitous and will be found on virtually every beach and riverbed in the region. The Gunflint Range of Minnesota and Ontario is an enormous chert formation containing iron minerals and stromatolite fossils, making the Grand Marais, Minnesota, and Thunder Bay areas lucrative.

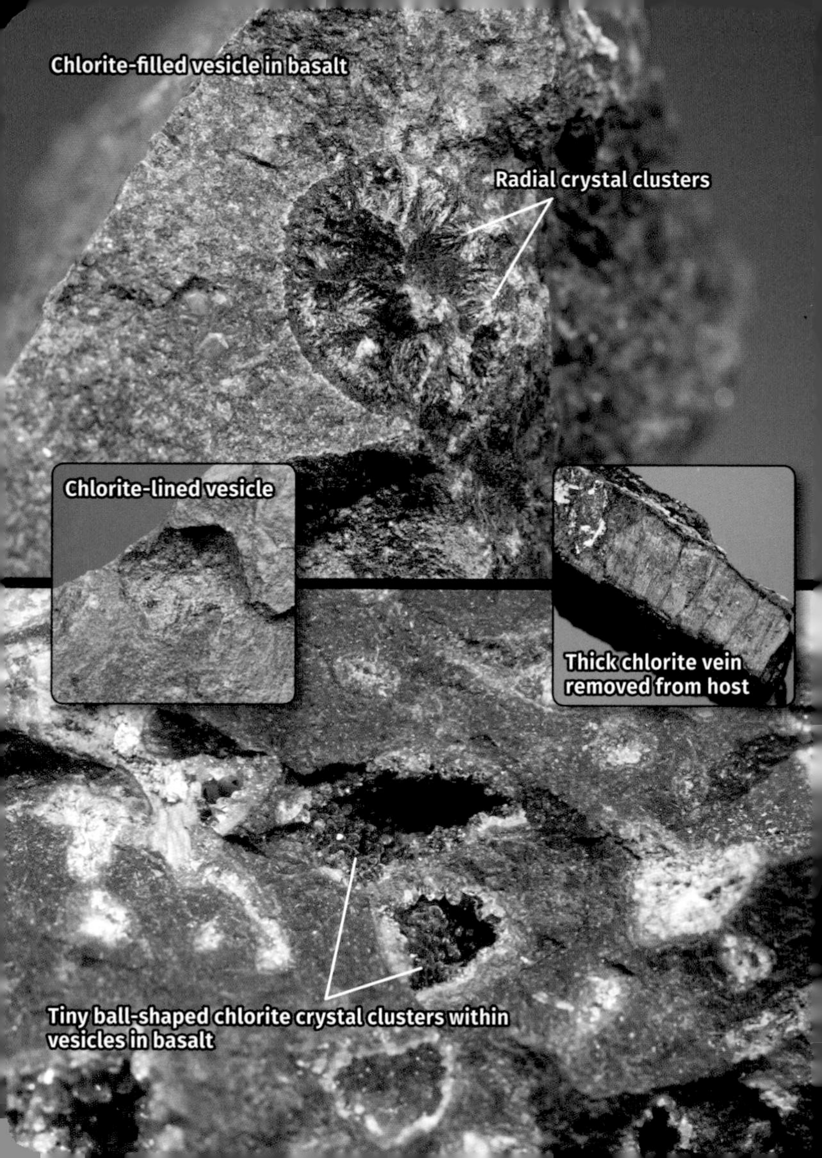

Chlorite-filled vesicle in basalt

Radial crystal clusters

Chlorite-lined vesicle

Thick chlorite vein removed from host

Tiny ball-shaped chlorite crystal clusters within vesicles in basalt

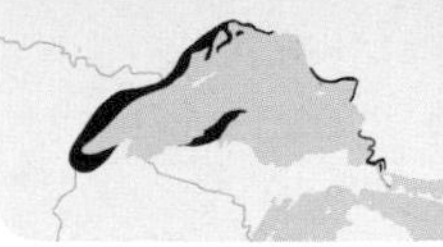
Primary Occurrence

Chlorite group

Hardness: 2–2.5 **Streak:** Colorless

Environment: Shores, rivers, mines, outcrops

What to Look for: Soft, dark green, "greasy" looking mineral found inside vesicles (gas bubbles) in basalt

Size: Individual crystals are very tiny, rarely more than a millimeter or so; masses and veins can be up to several inches

Color: Green to dark green, black; occasionally brownish

Occurrence: Very common

Notes: The chlorite group is a family of closely related minerals that share a number of similarities with both micas and clays. Only two chlorite minerals are prevalent in the Lake Superior region—chamosite and clinochlore—but telling them apart is nearly impossible outside of a lab; most collectors find "chlorite" a sufficient-enough label. The region's chlorite formed primarily in basalt when gases and hydrothermal water altered and affected the minerals in the rock. This caused new minerals to form in the vesicles (gas bubbles) and cracks in the basalt, and in this setting chlorite was one of the first minerals to form. Most amygdules or veins of dark green chlorite show some indication of a crystal structure, though it may be subtle and require magnification. This includes tiny, thin, plate-like crystals arranged into parallel stacks or into radial, fan-shaped groups; very rarely, "fuzzy" ball-shaped groups may be found. More nondescript coatings or masses may be identified by their very low hardness and "greasy" luster and feel; it can also be found coating whole, freshly exposed agates that have not yet been weathered enough for the soft chlorite to be worn away.

Where to Look: You'll find chlorite anywhere you find basalt (and to a lesser extent, diabase), making Minnesota's shores one of the key areas for specimens. The Keweenaw Peninsula's mine dumps and outcrops also produce fine specimens, including crystals, as well as in the Thunder Bay area.

Malachite (green)
Chrysocolla vein (blue)
Botryoidal chrysocolla growths
Chrysocolla on copper
Dusty chrysocolla coating (blue-green) on rock

Chrysocolla

Hardness: 2–4 **Streak:** White to pale blue

Primary Occurrence

Environment: Mines, rivers, outcrops

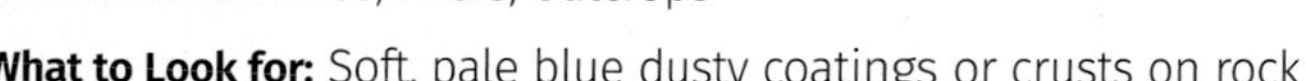

What to Look for: Soft, pale blue dusty coatings or crusts on rock or copper

Size: Coatings of chrysocolla are generally smaller than your palm

Color: Pale blue to greenish blue; rarely deep blue

Occurrence: Uncommon

Notes: Chrysocolla is a bluish green mineral that forms secondarily in copper deposits, which means that it forms after other chemicals have altered the copper and combined with it. As a result, it is usually found as a coating on copper or alongside other copper-bearing minerals, such as malachite (page 175), which formed in a similar manner. Chrysocolla never forms crystals, instead typically appearing as dusty crusts or masses; at best, it can exhibit botryoidal (bubbly, grape-like) masses clustered together. Chrysocolla also has a habit of drying out upon exposure to the air, making most specimens crumbly and chalky and easy to scratch away with a fingernail. While its color and occurrence on or near copper is very similar to that of malachite, malachite is usually more green and will exhibit a fibrous, sometimes velvety, texture or luster derived from its crystal structure. So while most specimens of chrysocolla are fairly unremarkable, the exception is "gem chrysocolla," a collectors' term for exceptionally colorful, solid, translucent masses that have been indurated (hardened) when silica solutions formed quartz around and within a chrysocolla mass. In all cases, the presence of chrysocolla can be used as an indicator of nearby copper.

Where to Look: Like all copper-bearing minerals, your best bet is the mine dumps and outcrops of the Keweenaw Peninsula. Lesser amounts can be found around Knife River and Beaver Bay, Minnesota, and Batchawana Bay, Ontario.

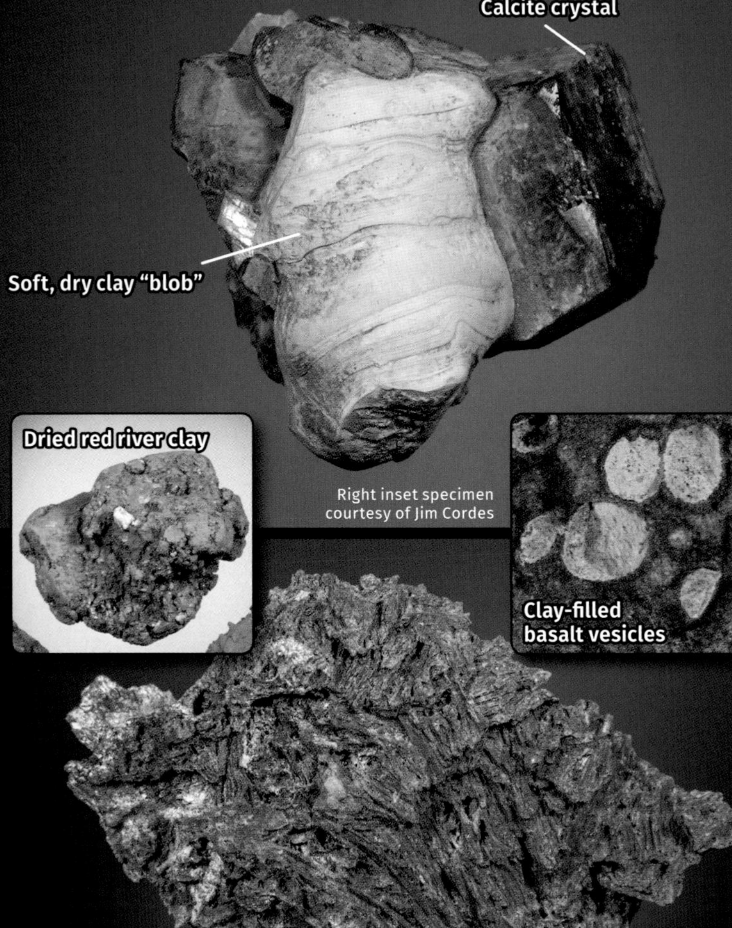
Calcite crystal
Soft, dry clay "blob"
Dried red river clay
Right inset specimen courtesy of Jim Cordes
Clay-filled basalt vesicles
Indurated clay with impressions from other minerals

Clay minerals

Primary Occurrence

Hardness: ~1–2 (higher when indurated) **Streak:** N/A

Environment: All environments

What to Look for: Very soft masses of material that easily crumble and have an earthy or chalky feel and texture

Size: Masses of clay can be virtually any size

Color: White to gray, yellow to brown, commonly reddish

Occurrence: Very common

Notes: Every rock hound has encountered sticky clay when searching wet riverbeds or muddy lakeshore outcrops, but few people consider what that clay actually is. Within a mass of clay are innumerable microscopic crystals of clay minerals, which are very soft, aluminum-bearing minerals that most often form from the chemical remains of weathered feldspars. Clay minerals—such as the very common kaolinite, illite, montmorillonite, smectite, and dickite—form as tiny stacks of flat, plate-like crystals that can trap water, causing the stacks to swell in size and making the entire mass of clay malleable and sticky. In the Lake Superior region, bodies of clay are rarely, if ever, composed purely of clay minerals, and instead incorporate some glacially pulverized rocks and organic matter. In its most interesting form, dry clay can be found as "blobs" deposited in rock cavities by dripping water. Some clays became indurated, or hardened, by the later introduction of silica (quartz material), making them much harder and more solid. Some indurated clays may preserve the impressions of other minerals no longer present.

Where to Look: Clays are ubiquitous and found anywhere. Minnesota's rhyolite cliffs near Duluth have produced interesting clays, including indurated masses. Copper mine dumps in Michigan also yield rare, tiny, white "fuzzy" crystals of a clay mineral called saponite, hidden in basalt with epidote.

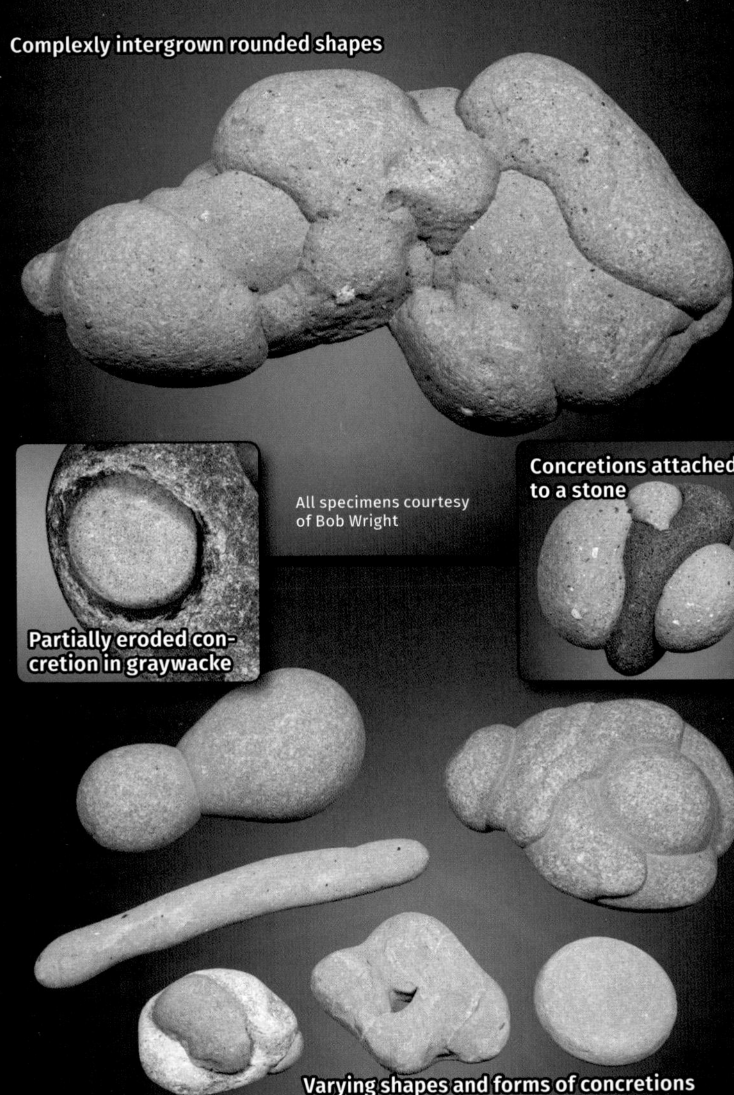
Complexly intergrown rounded shapes
Partially eroded concretion in graywacke
All specimens courtesy of Bob Wright
Concretions attached to a stone
Varying shapes and forms of concretions

Concretions

Hardness: N/A **Streak:** N/A

Primary Occurrence

Environment: Shores, rivers, outcrops

What to Look for: Conspicuously rounded, ovoid, or "blob" shaped formations found in or near sedimentary rocks

Size: Most are fist-size or smaller, but they can rarely be up to a foot

Color: Brown to tan or reddish, light to dark gray

Occurrence: Uncommon

Notes: Peculiar rocks with conspicuously rounded or globular shapes, concretions are among the most endearing rocks you can find in the region. They get their unique shapes not by weathering, as with most other rounded stones, but rather by how they formed. Concretions grow within developing sedimentary rocks that have been buried but not yet fully solidified. In this setting, minerals can nucleate, or collect, around a central point. In many cases, this can happen around an organic material, such as the remains of microbes, when carbon released from the decaying organics combines with minerals in groundwater to form new minerals, such as calcite. In other cases, a chemical reaction between minerals in the sediment and groundwater may cause mineral grains to accumulate upon each other as they harden. All concretions grow outward, locking the nearby sedimentary grains into a round mass. In most cases, concretions are of a different hardness than their host rock. Some are harder and are left behind after the host weathers away, as with northern Wisconsin's siltstone concretions. But others are softer and easily eroded, leaving a hole in their host rock, as in omarolluks (page 159). The most interesting and unusual of shapes—long "hot dogs" or conjoined "eggs"—are easy to identify as concretions because they could be little else; more difficult are the mundane finds, which will usually appear to just be water-worn rocks.

Where to Look: Northern Wisconsin is your best bet for concretions in the region. Exposed sedimentary rocks around Bayfield and Ashland produce many with wild shapes.

Close-up of rounded pebbles of varying sizes within conglomerate

Conglomerate & Breccia

Hardness: 6.5–7 **Streak:** White

Primary Occurrence

Environment: Shores, rivers, outcrops, pits

What to Look for: Rocks that appear to be made up of many smaller rocks that have been cemented or stuck together

Size: Both rocks can be found in any size

Color: Varies greatly; usually mottled and multicolored

Occurrence: Common

Notes: Conglomerate and breccia are two sedimentary rocks that formed as a result of smaller stones becoming cemented and hardened together into a solid mass. In conglomerate, the embedded stones are rounded, water-worn pebbles, cobbles, and sand that were deposited by ancient rivers, often where the rivers emptied into a larger body of water. The cement between the rounded stones varies; it can be a fine-grained sediment, but it is more often minerals like calcite, quartz, or goethite that were deposited by water and locked all the material together. Breccia is similar, but instead of rounded pebbles, it contains broken fragments of rock. When dramatic geological events like earthquakes or volcanic eruptions crush and break rock but don't scatter the fragments, mineral-rich water may inundate the material and cement it back together. Besides the shape of their constituent stones, another key difference between the two is that conglomerate may contain a mixture of pebbles of various kinds of rocks, while in breccia the fragments tend to be largely the same material.

Where to Look: You'll find conglomerate all around Lake Superior, but the tip of the Keweenaw Peninsula, near Copper Harbor, is home to the Copper Harbor Conglomerate, a large formation with extensive exposures along the shore. Breccia is also widespread, but is found more rarely.

Float copper coated in green malachite and black tenorite

Rounded edges

Weathered copper
in prehnite

Copper in
quartz (polished)

Ragged copper masses
partially freed of host rock

Chlorite (green)

Quartz (white)

Copper

Hardness: 2.5–3 **Streak:** Metallic reddish

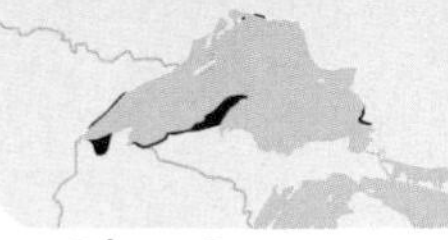

Primary Occurrence

Environment: All environments

What to Look for: Sheets or lumps of soft, metallic reddish metal, usually with a greenish or bluish surface coating

Size: Masses of copper may be fist-size and sometimes larger

Color: Metallic pinkish to orange-red when freshly exposed; dark gray, purplish brown, green to blue when weathered

Occurrence: Uncommon in Michigan, rare elsewhere

Notes: The Lake Superior region is famed for its world-class copper deposits. With a number of sources all around the lake, it has been extensively mined, beginning long before European settlers arrived. Copper is a native element; like gold or silver, a specimen of copper consists of just one element—copper—rather than a combination of elements as in most minerals. This makes it particularly easy to identify as its traits are uniform and consistent, no matter how it formed or where it's found. Its trademark red-orange metallic color is its most recognizable feature, but it only appears on freshly exposed pieces. Most of the time, copper will have a coating of other minerals derived from weathering and chemical alteration of the copper—namely green malachite (page 175), blue chrysocolla (page 113), black tenorite (page 219), and red to purplish brown cuprite (page 131)—but a fresh scratch will reveal the copper color below. Combined with its malleability and low hardness, there is very little you could confuse for copper.

Where to Look: The Keweenaw Peninsula is "copper country;" anywhere from gravel pits to riverbanks has potential, but mine dumps north of Houghton are the primary source. Knife River, Minnesota, is also known for copper, as is the Amnicon Falls area in Wisconsin, and Pancake Bay, Ontario.

Fine arborescent (tree-like) crystal
Crystal in datolite
Modified cubic crystal
Coarse, blocky crystal cluster
Arborescent crystal
Complex wiry crystals (dissolved from calcite)

Copper, crystalline

Hardness: 2.5–3 **Streak:** Metallic reddish

Primary Occurrence

Environment: Mines, outcrops

What to Look for: Blocky or plant-like growths of metallic reddish metal, usually with a greenish or bluish surface coating

Size: Copper crystals are rarely larger than three or four inches

Color: Metallic pinkish to orange-red when freshly exposed; dark gray, purplish brown, green to blue when weathered

Occurrence: Rare in Michigan; very rare elsewhere

Notes: Despite consisting of only one element, native elements like copper are still minerals, and as such, they crystallize in distinct shapes. The Lake Superior region is particularly well known for the fine copper crystals it has produced, and specimens can be found in museum collections all over the world. Copper has a cubic crystal system, which means that in their basest form copper crystals are cubes. But they rarely form so simply; most cubic crystals are highly modified, meaning that they have developed additional crystal faces or other variations, such as the elongation of certain edges. Crystals are typically found grown together in complex clusters, perhaps the best known of which are arborescent (tree- or plant-like) growths that branch outward from a single starting point. But there is quite a variation of shapes that can be seen in copper crystals; coarse, blocky points and complex tangles of angular wire-like growths (often dissolved out of masses of calcite) are also typical. All are rare, but they can be found in cavities (often in basalt) within or alongside prehnite, calcite, quartz, and rarely datolite.

Where to Look: Copper crystals are primarily only found in Michigan's mine dumps, but few survived the mining process. Any copper mine dump or waste rock pile, especially in the Houghton area and northward, has potential.

Copper "chisel chips"
Ridges formed by hammering
Electrolytic copper
Copper in slag
Copper-filled fire-brick
Copper pebbles from stamping mill

Copper, man-made

Hardness: 2.5–3 **Streak:** Metallic reddish

Primary Occurrence

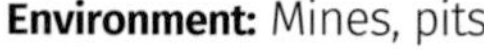

Environment: Mines, pits

What to Look for: Material exhibiting the properties of copper but in shapes or forms that don't match natural copper

Size: Man-made forms of copper may be up to several feet in size

Color: Metallic pinkish to orange-red when freshly exposed; dark gray, purplish brown, green to blue when weathered

Occurrence: Uncommon in Michigan; rare elsewhere

Notes: With such a long history of copper mining in the Lake Superior region, there are a number of ways in which the industry has left behind its own copper relics. Commonly confused for natural varieties of copper, man-made forms take on a number of appearances depending on how they were produced. "Chisel chips" are one example; produced a century ago when underground miners used a hammer and chisel to cut through enormous masses of copper, these elongated copper scraps have a uniform width and numerous ridges formed by hammering. At ore-processing sites, chimneys and other brick structures at smelting (ore melting) facilities were in contact with copper-bearing fluids and gases so long that the pores in the brick became filled with copper and copper-bearing minerals. At ore-crushing facilities, called stamping mills, copper-bearing rocks were crushed and the copper itself pounded into nuggets that are easily confused with float copper (page 127). Rough, "frothy," copper-rich slag (waste material), usually with attached bits of rock and glass, and lumpy, botryoidal (grape-like) electrolytic copper—a byproduct of copper plating processes—can also be found at old ore-processing sites.

Where to Look: Michigan mine dumps in the Houghton-Hancock area have produced chisel chips, and the nearby sites of former mining facilities may still yield bits of copper fire-brick. Stamped copper nuggets still turn up at mine sites.

Float copper coated in green malachite and red cuprite
Copper within calcite
Sheet copper in shale
Small copper veins (dark colored) in conglomerate

Copper, varieties

Hardness: 2.5–3 **Streak:** Metallic reddish

Primary Occurrence

Environment: All environments

What to Look for: Sheets or lumps of soft, metallic reddish metal, usually with a greenish or bluish surface coating

Size: Masses of copper may be fist-size and sometimes larger

Color: Metallic pinkish to orange-red when freshly exposed; dark gray, purplish brown, green to blue when weathered

Occurrence: Uncommon in Michigan, rare elsewhere

Notes: Most of the region's copper formed in cavities and fissures in extant overlying rocks when hydrothermal water (hot, mineral-rich groundwater and steam) rose upward and deposited minerals in it. As such, the area's copper can be found in a wide variety of settings, from vesicles (gas bubbles) in basalt and the paper-thin cracks in shale to the gaps between pebbles in conglomerate. A common occurrence is as irregular, gnarled masses tightly intergrown with epidote and quartz; in general, these dense, often sharp pieces of rocky copper ore are "leftovers" from mining operations. Much rarer and more desirable are crystals of calcite containing perfectly preserved, brightly lustrous copper encased within. But among the most common ways you'll find copper is as "float." Float copper is copper that was freed of its host rock by the crushing weight of the glaciers and then transported, or "floated," across the region within them. The ice rounded and weathered the copper, creating smooth nuggets or thick sheets, usually coated in green malachite.

Where to Look: Rivers along Minnesota's Superior shores and in Michigan's Keweenaw Peninsula can yield float copper; a waterproof metal detector is a great help. Copper-bearing conglomerate can be found around Copper Harbor, Michigan, and Batchawana Bay, Ontario.

Cleaned specimen in host rock
Copper
Silver
Epidote-rich basalt host
Stamped specimen
Rough specimen
Crude silver crystals
Copper mass
Cleaned specimen

Copper-Silver combination

Hardness: 2.5–3 **Streak:** Varies

Primary Occurrence

Environment: Mines, outcrops, rivers

What to Look for: Soft, metallic masses of both copper orange and silvery gray color, usually with dark surface coatings

Size: Masses are generally no larger than a few inches

Color: Metallic pinkish to orange-red and silvery gray; usually with green and/or gray surface coatings

Occurrence: Rare in Michigan; very rare elsewhere

Notes: The Lake Superior region's native copper (page 121) and native silver (page 213) formed as a result of the same hydrothermal activity (hot, mineral-rich groundwater that rose and deposited minerals in overlying rocks), so it should come as no surprise that both metals can occur together. Referred to by miners long ago as "copper-silver halfbreeds," specimens of natural copper-silver combinations exhibit small masses or crystals of silver grown atop a larger mass of copper (very rarely the opposite). While generally regarded as quite rare, they may actually be more common than we realize; when a specimen is coated in the crust of minerals (malachite, etc.) that naturally develop on its weathered surfaces, any silver that may be present is often dark, dull, and easily overlooked. But when these specimens are chemically cleaned, stripped of their superficial coating of minerals, the color differences in the copper and silver become apparent. Small nuggets may be found as well and usually lack any surface-coating minerals. These rounded, smoothed specimens have been through a stamping mill, their host rock crushed away and the metals mashed together.

Where to Look: Keweenaw Peninsula mine dumps, around Houghton and northward, are the primary place—and generally the only place—to find copper-silver combinations. They may very rarely turn up in Minnesota's and Ontario's copper deposits.

Float copper with malachite (green) and cuprite (dark red) coating

Gemmy red cuprite on copper in feldspar

Cuprite (orange-red) on copper crystal

Rare mass of fine translucent cuprite

Specimen courtesy of Alex Fagotti

Cuprite

Hardness: 3.5–4 **Streak:** Brown-red

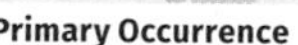

Primary Occurrence

Environment: Mines, rivers, pits, outcrops

What to Look for: Dark red crusts or masses in close association with copper and copper-bearing minerals, like malachite

Size: Individual crystals are generally tiny; masses and crusts are usually up to several inches in size

Color: Deep red to brownish red, purple-red, orange-red

Occurrence: Uncommon

Notes: While a majority of copper-bearing minerals are green, blue, or black in color, cuprite breaks the trend with its distinctive deep-red hues. A simple combination of copper and oxygen, it forms virtually anywhere copper is found, usually directly upon native copper (page 121), and it is closely associated with malachite, chrysocolla, and other minerals that form when copper weathers. Cuprite crystals are rare and usually very small, exhibiting cubic or octahedral shapes and often gem-like translucency and bright luster. But in the Lake Superior region you'll typically only find cuprite as crusts or coatings that are often fairly dull in appearance. Upon closer inspection, however, many cuprite crusts may appear to have tiny spots of "glitter"—those are actually tiny crystal faces. At best, its color is deep blood-red, but many specimens will be more brown and less commonly orange-red, especially when found underwater. Due to its occurrence with copper, it's unlikely you'll confuse it with anything else; it may appear superficially similar to some garnets (page 151), but it is much softer.

Where to Look: You'll find some amount of cuprite in any copper-rich area, the most productive of which are the many mine dumps in the Keweenaw Peninsula. The Mamainse Point area of Ontario and, to a lesser extent, the Knife River area in Minnesota also yield copper-bearing minerals.

Michigan datolites
Polished examples
Whole, unbroken nodules
Polished examples
Datolite crystals
Datolite nodule in prehnite
Water-worn nodules
Whole nodules
Polished examples
Minnesota datolites

Datolite

Hardness: 5–5.5 **Streak:** White

Primary Occurrence

Environment: Shores, rivers, mines

What to Look for: Rough, gray to white nodules with lumpy exteriors and smoother, fine-grained interiors

Size: Datolite nodules are typically thumbnail-size, but they can rarely be much larger, up to 6 or 8 inches in size

Color: White to gray or brown exteriors; white to gray, pink to red, yellow to brown, and greenish to blue (rare) interiors

Occurrence: Uncommon to rare; Michigan and Minnesota only

Notes: Once only known from Michigan but now also found at numerous localities in Minnesota, the datolite nodules found in the Lake Superior region are unique. Elsewhere in the world, datolite is found only as glassy crystals, but here it instead develops in cavities as clusters of tightly compacted microscopic grains. The exteriors of these generally round masses are usually lumpy and rough, often described as "cauliflower-like." But when broken or cut open, their interiors exhibit a smoother, more porcelain-like texture. Most are white or gray inside, but prized examples are tinted by mineral impurities in shades of pink, red, blue, and yellow. Chalky, water-worn datolites are found on beaches and rivers; masses of calcite can appear similar, but are softer and more crumbly. Freshly exposed datolites still in their host rock are often found with prehnite in muddy pockets containing multiple nodules. At some Michigan mine sites, small glassy, clear crystals can rarely be found and may resemble analcime (page 227), but datolite crystals are usually blockier.

Where to Look: Lake Superior shoreline and adjoining rivers between Duluth and Silver Bay, Minnesota, yield specimens. In Michigan, many of the Keweenaw's beaches are fruitful, especially near Keweenaw Point at the tip of the peninsula.

Beach-worn diabase
Freshly broken surface
Freshly broken texture
Weathered texture
Beach-worn diabase
Thomsonite

Diabase

Hardness: >5.5 **Streak:** N/A

Primary Occurrence

Environment: Shores, outcrops, rivers, pits

What to Look for: Dense, dark rock with a visibly grainy appearance and often with light-colored flecks throughout

Size: Can be found in any size, from pebbles to boulders

Color: Dark gray to black, occasionally greenish to brown; usually slightly mottled with lighter spots

Occurrence: Common

Notes: Diabase is an igneous rock of intermediate grain size between basalt (page 97) and gabbro (page 149). All three kinds of rocks have essentially the same mineral composition of plagioclase feldspars, pyroxene minerals, and olivine but differ in how large or well crystallized their component minerals have grown. Gabbro, having cooled slowly deep in the earth, has large, easily visible mineral grains, while basalt, which cooled rapidly on the earth's surface, has very small, fine grains. Diabase formed at a shallow depth and developed mineral grains that are still much finer than those of gabbro, but noticeably larger than those of basalt (this is best observed under magnification). The result is a dark, dense, moderately fine-grained rock with a greenish tint that appears very similar to basalt until inspected closely. When weathered and rounded, it becomes easier to see how it differs from basalt; beach pebbles often show poorly delineated light gray flecks of feldspars that give the stone a spotted appearance. And while it is quite common, you'll find diabase more sparingly than basalt on most shores. Lastly, voids in diabase may host other minerals, like zeolites.

Where to Look: Diabase is largely limited to the western end of Lake Superior, especially along Minnesota's shores; some of the largest exposures in the area are hills composed of diabase in Duluth.

Epidote (green) crusts on basalt
Water-worn sample
Well-formed crystals
Epidote crystals in basalt vesicle
Epidote (green) crusts on host rocks
Unakite pebble

Epidote

Hardness: 6 **Streak:** Colorless to gray

Primary Occurrence

Environment: All environments

What to Look for: Hard, yellow-green crusts or masses on rock, sometimes with elongated, tablet-like crystals

Size: Crystals are typically smaller than ¼ inch; crusts or masses may be several inches and rarely larger in size

Color: Yellow-green color is distinctive; it can also be dark green

Occurrence: Common

Notes: There are a number of green minerals you'll find along Lake Superior's shores, but few are as easy to identify as epidote. Formed when calcium-rich feldspars within rocks are altered by external forces, epidote is usually found within cavities and cracks in basalt and some granitoids. Crystals are common and often attractive, appearing as small tablet- or blade-like plates, sometimes arranged into radiating or fan-shaped groups and generally exhibiting bright glassy luster. Many examples will also show deeply striated, or grooved, crystal faces. But as abundant as crystals may be, crusts and veins are still far more common, and they usually don't bear any indication of crystal structure. Instead, these can often be identified by color alone. Color is typically a poor identifying trait for minerals, but epidote's yellow-green or "pistachio green" coloration is so distinctive that it will be your first clue in identification. Combined with its hardness and frequent association with quartz and feldspars, you shouldn't have any trouble. Unakite (page 221), a variety of granite, is an abundant epidote-bearing collectible as well.

Where to Look: Pebbles of rocks containing epidote, such as unakite, are common on virtually any shoreline, but especially in Ontario. Crystals are best found in vesicles within basalt in the Keweenaw Peninsula and all along Minnesota's shores.

Granite with tan to pink feldspars

Granitoid with red feldspars

Feldspar in syenite

Water-worn feldspar masses

Microcline in basalt vesicle

Granite with elongated feldspar crystal showing schiller effect

Feldspar group

Hardness: 6–6.5 **Streak:** White

Primary Occurrence

Environment: All environments

What to Look for: Abundant, hard, light-colored masses or blocky crystals embedded in rocks or found inside vesicles

Size: Varies; masses and crystals, both loose and embedded in rocks, typically range from sand-size up to an inch or two

Color: White to gray, tan to brown, orange to pink or red

Occurrence: Very common

Notes: Feldspars are everywhere; they make up nearly 60 percent of the earth's crust and can be found as constituent minerals in most types of rocks. But "feldspar" is a general name for an entire group of closely related minerals, which are further subdivided into the potassium feldspar group (or "K-spars"), including the very common orthoclase and microcline, and the plagioclase feldspar group, including albite and anorthite. The difference between the two subgroups is largely chemical; for rock hounds, the general guideline is that K-spars are abundant in light-colored rocks like granite and rhyolite, usually as small embedded grains or blocky masses, while plagioclase feldspars are often found in darker rocks, like gabbro. When embedded in rocks, masses of feldspars can be identified by their hardness, generally glassy luster, and their common habit of reflecting light with an internal schiller, or "flashiness." Embedded crystals are often longer than they are wide, while free-standing crystals (called "adularia," see page 141) are blocky or wedge-shaped. Water-worn masses can also be found on shores, freed from their Canadian host rock by the glaciers.

Where to Look: You can find water-worn feldspars on virtually any stretch of Lake Superior shoreline, especially along Ontario's shores as they're closer to the Canadian Shield.

Pumpellyite (gray-green)

Epidote (dark green)

Adularia crystals (orange)

Adularia and epidote in calcite

Adularia with calcite and copper

Adularia crystals lining a vesicle in basalt

Feldspar, "Adularia"

Hardness: 6–6.5 **Streak:** White

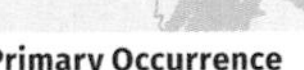

Primary Occurrence

Environment: All environments

What to Look for: Small orange wedge-shaped crystals, usually within vesicles in basalt, often alongside epidote or calcite

Size: Individual crystals are usually smaller than ¼ inch; masses or amygdules may be up to a few inches in size

Color: Orange to pink or salmon-colored, also tan; rarely white

Occurrence: Common

Notes: The name "adularia" doesn't refer to one specific mineral but rather to a potassium feldspar (page 139) that has formed in such an environment that sharp, free-standing crystals were able to develop (as opposed to the embedded masses usually seen in rocks like granite). In the Lake Superior region, most "adularia" tested in a lab is found to be microcline, developed in low-temperature hydrothermal settings where mineral-rich water left behind crusts or masses of crystals. This means you'll generally find adularia within vesicles and cavities in basalt and other volcanic rocks where its typical orange or salmon-pink coloration is distinctive. By its nature, it is usually found nicely crystallized; crystals appear as sharp wedges, and if complete will exhibit a rhombohedral cross-section, shaped like a "leaning cube." Crystals are often tightly intergrown with each other in complex groups that can line the walls of a cavity completely. These traits, combined with its high hardness, generally dull (though sometimes glassy) luster, and its common association with epidote and calcite make adularia easy to identify.

Where to Look: Adularia is widespread but you'll mostly easily find it in basalt of the Keweenaw Peninsula of Michigan. Mine dumps and outcrops there yield countless specimens.

reen fluorite
conglomerate
lichigan)
Purple fluorite
vein in rhyolite
(Minnesota)
Water-worn samples
(Minnesota)
Masses broken free
of vein (Minnesota)
Cubic crystals on
amethyst (Ontario)
Cubic crystals
(Ontario)
Ontario fluorite
arge pale purple cubic
ystals on quartz
Well-formed, intergrown cubic
crystals on granite

Fluorite

Hardness: 4 **Streak:** White

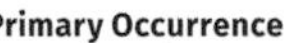

Primary Occurrence

Environment: Shores, rivers, mines, outcrops

What to Look for: Glassy purple or green cubic crystals, crusts, veins or masses of low hardness and exhibit angular breaks

Size: Individual crystals are rarely larger than your thumbnail; crusts or masses may be several inches to softball-size

Color: Pale to dark purple or blue; pale to dark green; yellowish

Occurrence: Uncommon; fine crystals are rare

Notes: The Lake Superior region is not particularly well known for fluorite, yet a surprising number of fine specimens are found all across the area. This fluorine-bearing mineral turns up in a variety of Lake Superior's rocks in shades of green and purple (less commonly yellow or bluish) and is usually glassy and translucent. Crystals are primarily cubic, often sharply developed and intergrown in complex groups that coat the surfaces of a cavity in rock. As with most minerals, however, poorly formed veins or masses are more common, and they would go overlooked if it weren't for their typically conspicuous coloration. In all cases, fluorite's hardness is distinctive, distinguishing it from purple amethyst (which is harder), and impure calcite (which is softer). You can also check for its perfect cleavage; fluorite, when carefully struck, will break at sharp triangular angles, leaving a smooth surface. And as a mineral usually deposited by hydrothermal water after an area's rocks formed, it can be found within a wide variety of materials, from granite to rhyolite to conglomerate.

Where to Look: Numerous fantastic Lake Superior localities exist: the Thunder Bay region, east of Thunder Bay, Ontario, is known for crystals on amethyst; the Copper Harbor Conglomerate in Michigan produces green crystals; and Minnesota's shore near Duluth yields veins of fluorite in rhyolite.

Undetermined fossil in chert

Embedded seashell fragments in chert

Coral fossil

Specimen courtesy of Daniel Bubalo, Jr.

Water-worn petrified wood

Coral fossil (Michigan)

Gauzy texture of coral in limestone

Horn coral fossils in limestone

Specimens courtesy of Bob Wright

Fossils

Hardness: N/A **Streak:** N/A

Primary Occurrence

Environment: Rivers, shores, pits

What to Look for: Rocks, especially limestone and chert, with shapes, impressions and indications of life-forms

Size: Typical specimens are usually no larger than your palm

Color: Tan to gray or brown, sometimes rust-colored

Occurrence: Rare

Notes: Fossils are the remains of ancient plants and animals that have been preserved within rock. They formed when organic matter, such as a branch or a clam shell, was buried in sediment underwater where a lack of air prevented it from decaying normally. Over the course of millions of years, minerals dissolved in groundwater interacted with the buried remains, crystallizing in and around them, resulting in a cast or impression that preserved the shape of the plant or animal. This process only occurs in sedimentary rocks, particularly in limestone, shale, and chert. You won't find a great variety of fossils in the Lake Superior region, largely because of the glaciers that scoured the area, but the region's aquatic past does mean that fossil corals can be found with some frequency. Corals exhibit a gauze-like texture or radiating patterns when worn, and cone- or cup-like shapes when well preserved. Petrified wood may also rarely turn up, deposited by the glaciers; look for a distinct, fine wood-grain texture in jasper. A quick examination of any chert finds may occasionally reveal unexpected aquatic fossils, too. But the area's most common fossils are stromatolites, discussed separately on page 147.

Where to Look: Fossils are generally scarce around Lake Superior, but glacially deposited corals and wood can turn up on most shores. The eastern half of Lake Superior hosts more sedimentary rocks and therefore more potential for fossils, especially the eastern shores of the Upper Peninsula.

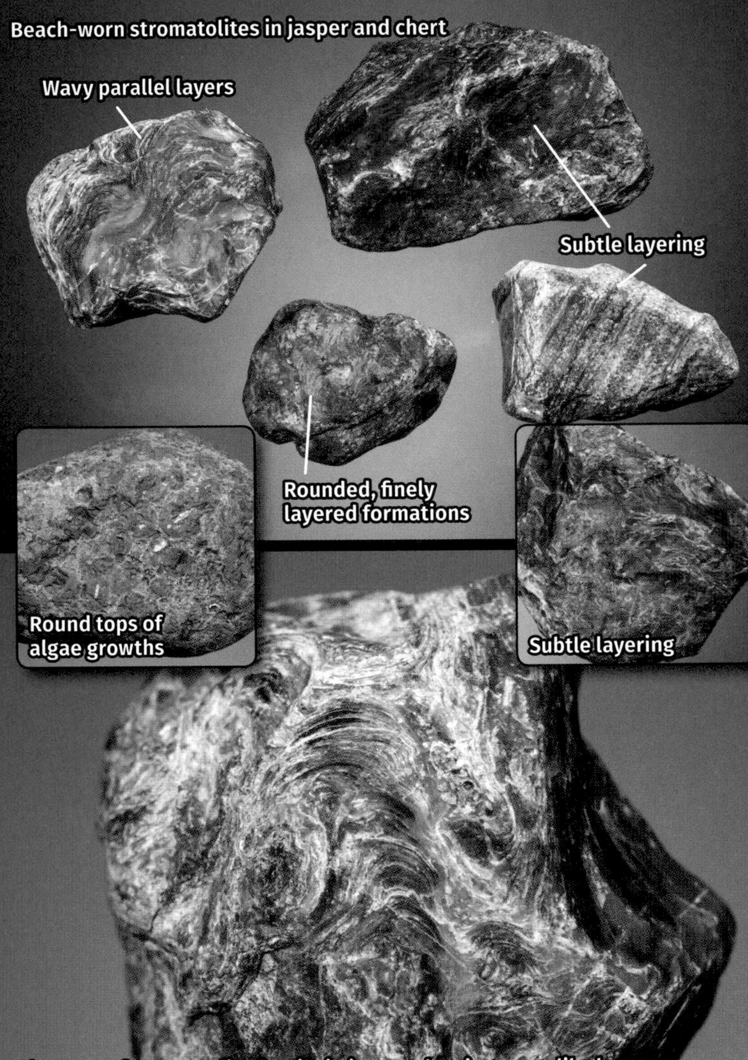
Beach-worn stromatolites in jasper and chert
Wavy parallel layers
Subtle layering
Subtle layering
Rounded, finely
layered formations
Round tops of
algae growths
Subtle layering
Close-up of stromatolite fossils in jasper showing wave-like layers

Fossils, stromatolite

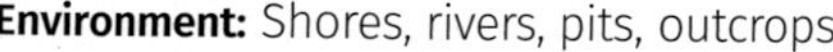

Hardness: >6.5 **Streak:** N/A

Primary Occurrence

Environment: Shores, rivers, pits, outcrops

What to Look for: Hard, opaque masses with tight parallel layering that has curving or wavy shapes

Size: Typical loose specimens are usually softball-size or smaller

Color: Varies; usually multicolored in gray to black, brown to red, yellow to orange, and more rarely green

Occurrence: Uncommon

Notes: Stromatolites are among the most important fossils on the planet, and Lake Superior happens to be home to many. Appearing as fine, wavy layers in chert and jasper, these seemingly mundane bands of color are actually fossilized cyanobacteria. Cyanobacteria, or blue-green algae, were one of the first life-forms on Earth, appearing in the early oceans around 3.6 billion years ago, and they are credited with turning Earth's harsh atmosphere into a more oxygen-rich one through photosynthesis. Still living in the oceans today, stromatolites appear as sticky mats that grow upward on top of themselves, creating tower- or mushroom-shape structures. Long ago, these became embedded in the silica-rich muds at the bottoms of seas, preserving them in hard chert and colorful jasper. Arch-like bands of color or wavy parallel lines are the key identifiers today, though you may have to look closely for them because they can be subtle. Banded iron formation (page 93) can look similar but is usually darker, and agates (page 39) are more translucent.

Where to Look: Stromatolite fossils can be found all over the region's shores, but any of Minnesota's Lake Superior beaches will be a good place to start. The Gunflint Range, largely in Ontario, is also home to extremely old and scientifically significant stromatolite formations, and the Thunder Bay area will also be worth searching.

Broken samples
Coarse texture
Close-up of texture
Weathered sample with rusty olivines
Weathered and water-worn samples

Gabbro

Hardness: >5.5 **Streak:** N/A

Primary Occurrence

Environment: Shores, rivers, pits, outcrops

What to Look for: Dark, coarsely grained rock containing glassy greenish and black grains

Size: Gabbro can occur in any size, from pebbles to entire cliffs

Color: Mottled gray to black, greenish to yellowish, brown

Occurrence: Common

Notes: Gabbro is a dark, coarse-grained igneous rock with the same plagioclase feldspar-, pyroxene-, and olivine-rich composition as basalt (page 97) and diabase (page 135), but it formed deeper in the earth, allowing its minerals to grow to a larger size. After the Midcontinent Rift began, around 1.1 billion years ago, the initial lava flows, composed largely of basalt, formed a rock "ceiling" that insulated the molten rock below it. The deeper rock cooled slowly and in such immense amounts that the resulting gabbro formations are some of the largest in the world. The crystals embedded in gabbro may be up to thumbnail-size and rarely larger, with glassy translucent feldspars and dark, reflective pyroxene minerals being prominent and easily spotted in freshly broken samples. Weathered, water-worn stones are often darker with less obvious crystals, but their green-black mottled appearance will still be distinctive. In general, its texture is very similar to that of granite, but gabbro is always darker. In very highly weathered examples, the constituent grains of olivine have largely turned into rust-colored iron oxides.

Where to Look: Most of the region's gabbro is found along Minnesota's Lake Superior shoreline; the Duluth Gabbro Complex, covering much of northeastern Minnesota, is among the largest gabbro formations in the world. Glacially deposited pebbles will be present many other places as well.

Cut gneiss cobble containing garnet crystals (red)

Specimen courtesy of Bob Wright

River sand garnet (approx. 1⁄16")

Andradite (yellow)

Right inset specimen courtesy of Jim Cordes

Granitoid beach cobble containing garnet crystals (red)

Specimen courtesy of Bradley A. Hansen

Garnet group

Hardness: 6.5–7.5 **Streak:** Colorless

Primary Occurrence

Environment: Shores, rivers, pits

What to Look for: Hard, rounded, usually reddish crystals embedded in coarse-grained rocks or loose on river bottoms

Size: Most garnets are under a half inch and usually smaller

Color: Red to brown; rarely yellow to orange-yellow

Occurrence: Uncommon to rare

Notes: Garnets are a common and often colorful gemstone that, with some patience, can be found in the Lake Superior region. But the name "garnet" doesn't refer to a single mineral but rather to a large group of closely related minerals with similar chemical compositions and crystal structures. Garnets all tend to be very hard, translucent and glassy, and—most distinctively—they form as small, round crystals resembling faceted balls when well formed, often with striated (grooved) faces. In this region, a number of garnet minerals are present but only almandine is abundant enough to find with any frequency. It usually only appears as conspicuous reddish angular spots embedded in granite, gneiss, and other coarse-grained rocks deposited by the glaciers. Almandine (or other similar red garnets) can also be found in sand as tiny, gemmy grains weathered from their host rock. Andradite is a much rarer garnet that formed in metamorphic conditions; tiny yellow crystal clusters can rarely be found within cavities in altered basalt, often with epidote.

Where to Look: Gemmy red garnets can be found in nearly any river's sand, if you use magnification and have patience to look for it. Almandine is found in water-worn granite cobbles, especially on Ontario's shores. And yellow andradite, while quite rare, can be found in mildly metamorphosed basalts in the Grand Marais, Minnesota, area.

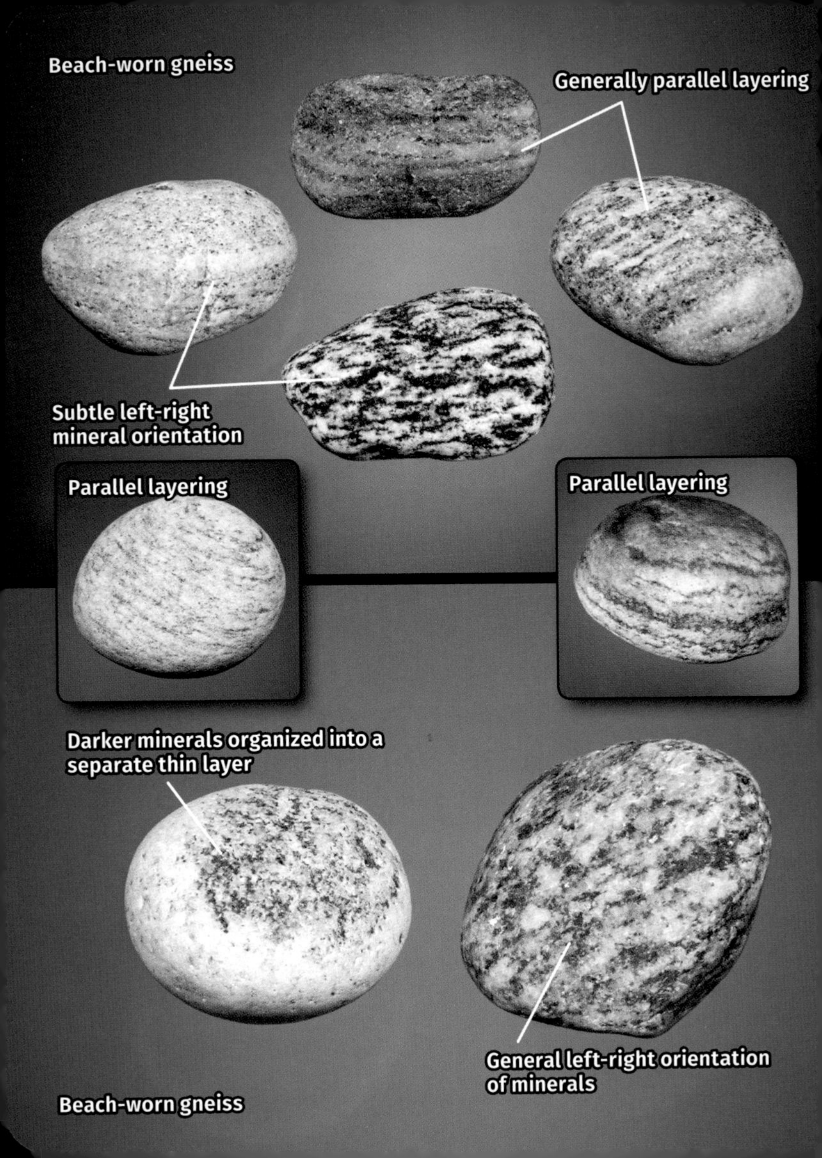
Beach-worn gneiss
Generally parallel layering
Subtle left-right mineral orientation
Parallel layering
Parallel layering
Darker minerals organized into a separate thin layer
General left-right orientation of minerals
Beach-worn gneiss

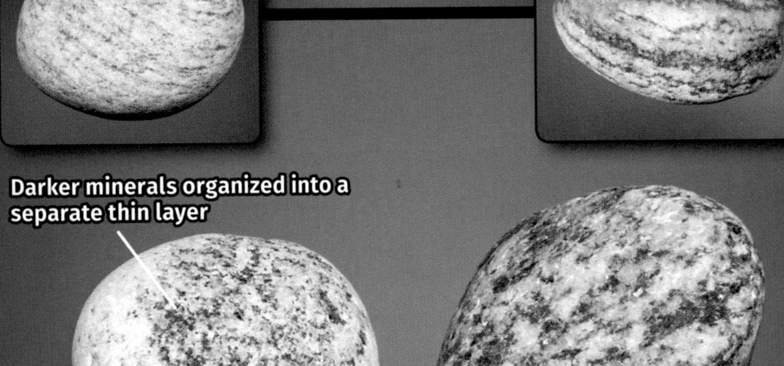

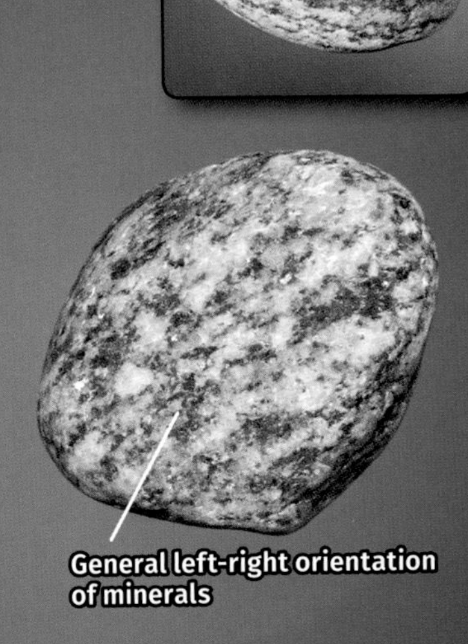

Gneiss

Hardness: N/A **Streak:** N/A

Primary Occurrence

Environment: Shores, rivers, pits

What to Look for: Hard, dense rocks of varied grain sizes arranged into often subtle parallel layers

Size: Gneiss can be found in any size, from pebbles to boulders

Color: Multicolored; mottled white to gray or black, brown to pink, tan, sometimes greenish or yellowish

Occurrence: Common

Notes: Gneiss (pronounced "nice") is a dense, layered metamorphic rock that forms when sedimentary and especially igneous rocks are subjected to high heat and high pressure deep within the earth. These extreme forces cause the minerals within the rock to soften and be more easily moved and deformed, a state called plasticity. As the hot, plastic rock is squeezed and pushed by pressure, it can shear, or slide side-to-side, which stretches or "smears" the structure of the rock. This process gives gneiss its characteristic layering, which is very coarse and subtle in lightly metamorphosed rocks but very pronounced and bold in heavily metamorphosed examples. The layers can be uniform and parallel, or wavy and of varying thickness, depending on the level of metamorphism. The metamorphic processes can also cause new minerals to develop in the rock, including gems like garnet (page 151). Most of the gneiss you'll find around Lake Superior is derived from granite or granitoids (page 157) originating from the Canadian Shield, and they will still resemble their parent rock. The most difficult specimens to identify are those with very subtle layering; look for dark, glittering mica flecks that are clustered into elongated groups.

Where to Look: Gneiss is a common shore find on any cobble beach; try the shores of Ontario and in northern Minnesota.

Goethite mass (dark brown) with limonite-coated surfaces (yellow)

Limonite on sandstone

Goethite

Chert colored by limonite

Limonite on chalcedony

Weathered pyrite (now limonite)

Agates with limonite surface coatings

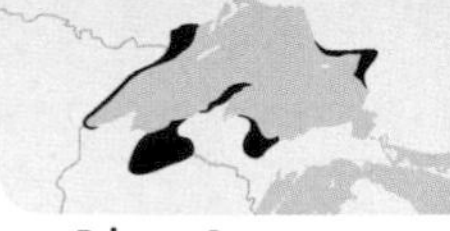

Goethite & Limonite

Hardness: ~4–5.5 **Streak:** Yellow-brown

Environment: All environments

What to Look for: Metallic brown masses with yellow surface coatings, or yellow-brown rusty surfaces on various rocks

Size: Masses are generally smaller than your fist

Color: Goethite is metallic black to brown when solid; more commonly yellow-brown to rusty orange-brown

Occurrence: Goethite uncommon; limonite very common

Notes: The Lake Superior region is rich with iron ores, and the most prominent are hematite (page 161) and goethite. Goethite (pronounced "gur-tite") is a hydrogen-bearing iron oxide favored by the mining industry and formed primarily as a result of the weathering and breakdown of other iron-bearing minerals. While common in the inland iron ranges, crystals or botryoidal (lumpy, grape-like) masses of goethite are typically not found on the shore. Instead, irregular chunks, often water-worn, may turn up sparingly and can be identified by goethite's typical brown or black metallic surfaces, usually stained a rusty yellow. Such masses will resemble hematite and magnetite, but hematite has a distinctly reddish streak while goethite's is yellowish, and goethite is not magnetic like magnetite. Far more common is limonite, a water-bearing mixture of iron oxides that forms primarily when other iron-bearing minerals weather and degrade; most people know it better as rust. Goethite is a primary component of limonite and it can most often be found coating other rocks and minerals, including on and within agates. Virtually anywhere you see a rusty yellow stain, limonite can be assumed present.

Where to Look: The Lake Superior region is home to large iron ore deposits, including goethite, but shore finds of goethite are fairly rare. Limonite, on the other hand, is widespread; you'll find it on agates, chert, and other rocks.

Beach-worn granite cobbles
Granitoids
Close-up of texture
Diorite
Rough granite
Granitoids

Primary Occurrence

Granite & Granitoids

Hardness: N/A **Streak:** N/A

Environment: Shores, rivers, pits, outcrops

What to Look for: Hard, coarse-grained rocks containing visible grains and crystals of different minerals

Size: As rocks, granite and granitoids can be found in any size

Color: Varies greatly; mottled spots of white to gray, black, pink, brown, red to orange, yellow, and green

Occurrence: Very common

Notes: Granite is one of the most common rocks on the planet, and is a coarse-grained igneous rock that formed deep in the earth where its magma cooled very slowly, allowing the minerals within it to grow to large, visible sizes. But granite contains specific proportions of certain minerals—predominantly quartz, feldspars, amphiboles, and micas—and if any of those minerals are present in different proportions, then it becomes a different kind of rock. But how do you tell granite apart from a similar coarse-grained rock that has less quartz and more amphiboles, for example? Generally, you don't—not without professional analysis. We call these granite-like rocks granitoids, and for collectors the distinctions between them are usually very subtle. True granite is mostly light-colored with whites, grays, and pinks, and a lower percentage of dark spots. Reflective micas and lustrous rectangular feldspars are often visible in granite. There are numerous granitoids around Lake Superior, too, but a few are most common: granodiorite, diorite (both contain more dark minerals than granite, with diorite often being predominantly black), and syenite (discussed on page 215). Granite and most granitoids are hard, tough rocks, and the majority found on Lake Superior's western and southern shores originated in Canada, transported by the glaciers.

Where to Look: Granitoids are found on most Lake Superior shores, but increase in abundance closer to the Canadian Shield.

Graywacke omarolluks
Weathered,
rounded cavities
Close-up of texture
Calcite veins
in graywacke
Freshly eroded,
sharp-edged cavities
Omarolluk
All specimens courtesy of Bob Wright

Graywacke & Omarolluks

Hardness: N/A **Streak:** N/A

Primary Occurrence

Environment: Shores, rivers

What to Look for: Dark gray, grainy, soft rocks that are usually water-worn; they may contain very rounded holes

Size: Graywacke cobbles are usually no larger than a foot or so

Color: Light to dark gray, occasionally greenish or bluish gray

Occurrence: Graywacke uncommon; omarolluks rare

Notes: Graywacke (pronounced "gray-wacky") is a sedimentary rock unlike most others in that it does not consist of sediments of a uniform size or organization pattern. Graywacke resulted from underwater landslides or turbulent currents that mixed sediments of varying sizes together. This included grains of sand (primarily quartz and feldspars) and small pieces of rock, but the primary component is clay, which cemented the material together. Under magnification, you'll be able to see the larger grains of sand embedded in the softer, finer grained gray clay. This texture is subtle, but it is a key characteristic, especially since water-worn graywacke can resemble basalt (which is harder and much more common). Omarolluks, called "omars" for short, are a highly collectible kind of graywacke that contain ball-shaped holes, often perfectly round and seemingly unnatural. In fact, many collectors have mistaken omarolluks for manmade artifacts. But these conspicuous cavities formed when softer, embedded concretions (page 117) weathered out of the rock. These may, at first glance, look like vesicles (gas bubbles) in basalt, but are much more round and far fewer in number.

Where to Look: By tracing the path of the glaciers, it has been determined that omars originated in rock of the Hudson Bay area, Canada. Omars, as well as common graywacke, were deposited in the Lake Superior region by the glaciers and are found anywhere, though Wisconsin's shores are lucrative. Larger masses of graywacke may be found in the area west of Duluth, Minnesota.

Rough, weathered hematite
Rusty coating
Freshly exposed hematite
Granular hematite
Hematite vein
Prehnite

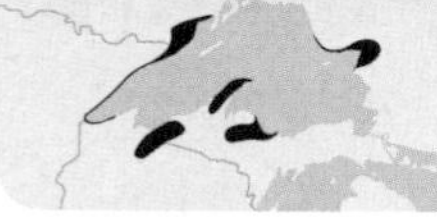

Primary Occurrence

Hematite

Hardness: 5–6 **Streak:** Reddish brown

Environment: All environments

What to Look for: Dark gray metallic masses or veins, usually with a red-brown surface coating

Size: Masses and crusts can vary greatly, from inches to feet

Color: Steel-gray to black; usually stained red to red-brown

Occurrence: Common

Notes: One of the most common iron-bearing minerals on the planet, hematite is a simple iron oxide mined extensively in the wider Lake Superior region. Much like goethite (page 155), it typically forms as dark brown or black metallic masses, often with a botryoidal (lumpy, grape-like) texture and fibrous cross-section, but it has a distinctly reddish-brown streak as opposed to goethite's yellow streak. In fact, most hematite specimens will be found with a rusty red-brown powdery coating or crust, which will aid in identification. Its lack of magnetism will also distinguish it from magnetite (page 173). Crystals are typically tiny, lustrous intergrown plates in rock cavities, but they're quite rare in any near-shore rocks. Instead, masses and veins with no distinctive shape and a grainy texture are the norm in the region. That's because Lake Superior's hematite was deposited as tiny grains settling into thick beds on the shores of ancient seas as early as 2.5 billion years ago, usually with chert, and later compressed and solidified into huge solid masses as seen in a banded iron formation (page 93).

Where to Look: While metallic chunks are not typical shore finds, they still may turn up occasionally on Minnesota and Michigan's shores. More abundant are veins or masses in rocks like basalt and banded iron formations all over the region, especially near the Gunflint and Gogebic ranges.

Close-up of gabbro

Plagioclase (gray)

Ilmenite grains (metallic black)

Ilmenite grains from river sand (largest approx. 1⁄16")

Ilmenite

Hardness: 5–6 **Streak:** Brown-black

Primary Occurrence

Environment: Shores, rivers, outcrops, pits

What to Look for: Small, weakly magnetic, metallic black grains embedded in dark rocks or found loose in black sand

Size: Most will be pea-size or smaller; usually found as tiny grains

Color: Metallic black to bluish black; may be dark brown to tan

Occurrence: Common

Notes: Ilmenite is an abundant titanium-bearing mineral, most common as a constituent in dark, iron-rich igneous rocks like basalt, diabase, and gabbro. Elsewhere, ilmenite can form attractive plate-like crystals, but in the Lake Superior region it is present virtually only as tiny, nondescript dark grains tightly embedded in their host rock. Large enough specimens can appear slightly bluish, and when highly weathered it can develop a dull tan to brown surface coating, but in general it is always brightly lustrous metallic black and opaque. It is also magnetic, and in combination with its appearance and occurrence in dark rocks, it is extremely easy to confuse with magnetite (page 173). But while magnetite is strongly magnetic, bonding tightly with a magnet, ilmenite is only weakly so, and while it will be attracted to a magnet, it can easily be shaken free. In addition, ilmenite has a conchoidal fracture (when struck, it breaks in a curved shape) and magnetite does not, but this won't be easily observable in most specimens. One of the simplest ways to find ilmenite will be to drag a magnet through black sand, as weathering often frees the dense metallic grains.

Where to Look: Ilmenite grains can be found as minuscule grains in any basalt, but for more visible grains you can search in the gabbro found along Minnesota's shores. River sand and black sand beaches will produce countless grains; try those along Sibley Peninsula, Ontario.

each-worn jasper

Common opaque, purple-red color

Stromatolite-bearing specime

Banded iron formation

Left inset specimen courtesy of Dave Woerheide

Beach-worn jasper

Right inset specimen courtesy of David Gredzens

asper mass with subtle layering

Specimen courtesy of Jim Corde

Jasper

Hardness: 7 **Streak:** White

Primary Occurrence

Environment: Shores, rivers, pits, outcrops

What to Look for: Hard, reddish stones with a waxy feel and appearance, often with visible layers of color

Size: Masses can be found in any size, from pebbles to boulders

Color: Generally red to red-brown, deep red to purple; also yellow to green

Occurrence: Very common

Notes: Jasper is a popular, colorful collectible that is abundantly available all around Lake Superior. While it forms in a variety of ways, it is composed of microscopic grains of quartz and therefore is considered a variety of chert (page 109). Typically defined by its colorful nature—most commonly tinted shades of red and brown by iron-bearing minerals—you'll find jasper with a variety of different textures and habits. These include being layered with dark hematite (called banded iron formation, discussed on page 93), common purplish water-worn pebbles with a granular appearance, and round nodules formed in cavities in rock. As a variety of quartz, jasper shares most of its traits: conchoidal fracture (when struck, circular cracks appear), high hardness, waxy luster when weathered, and sharp edges when broken. Jasper is easy to confuse with chalcedony (page 103), but it is opaque while chalcedony is translucent. Jasper is also frequently mistaken for agate (page 39), especially when it contains layers or bands of color, but agates are also translucent and typically have concentric ring-like banding.

Where to Look: You will find jasper virtually anywhere in the Lake Superior region, especially on Minnesota's shores near Grand Marais, on Michigan's Keweenaw Peninsula's shores, and Rossport, Ontario's area shores.

Common beach junk
Sheet steel
Aluminum
Beach glass
Glazed tile
Slag
Close-up of slag
Concrete
Brick
Slag
Asphalt
Wood

Junk

Hardness: N/A **Streak:** N/A

Primary Occurrence

Environment: All environments

What to Look for: Unnatural materials or objects that may mimic rocks or minerals; often appearing out of place

Size: Junk can be found in any size

Color: Varies greatly

Occurrence: Very common

Notes: When walking Lake Superior's shores in search of treasures, you will also invariably find trash. Most manmade junk will be obvious, but some materials can mimic naturally occurring rocks and minerals. A weathered and water-worn piece of concrete, for example, can greatly resemble conglomerate or breccia (page 119), but it is generally much harder and more solid. Beach glass, the rounded remnants of glass bottles or electrical insulators, can initially appear to be quartz or other collectible minerals, but is generally too clear and too uniform in thickness to be natural. Man-made metals can be found, too; rusty steel or iron sheets or machine parts may turn up, but their malleability and magnetism will distinguish them from anything natural. Melted soda cans turn up on shorelines as shiny aluminum "blobs," occasionally perplexing novices, but the biggest culprit for fooling collectors is slag. Slag is a glassy, bubble-filled byproduct of the mining industry. It may resemble basalt or rhyolite, but its bubbles and texture are too uniform. If a highly weathered piece leaves you in doubt, a fresh break will reveal slag's typical glassy or slightly metallic luster.

Where to Look: Wherever people have gone, junk will follow. Even the most remote shores of Lake Superior will bear some amount of man-made junk, from beach glass to concrete or fishing gear. The region's rich mining history means slag is common around many shoreline towns and especially railroads, as are bits of metal and tar.

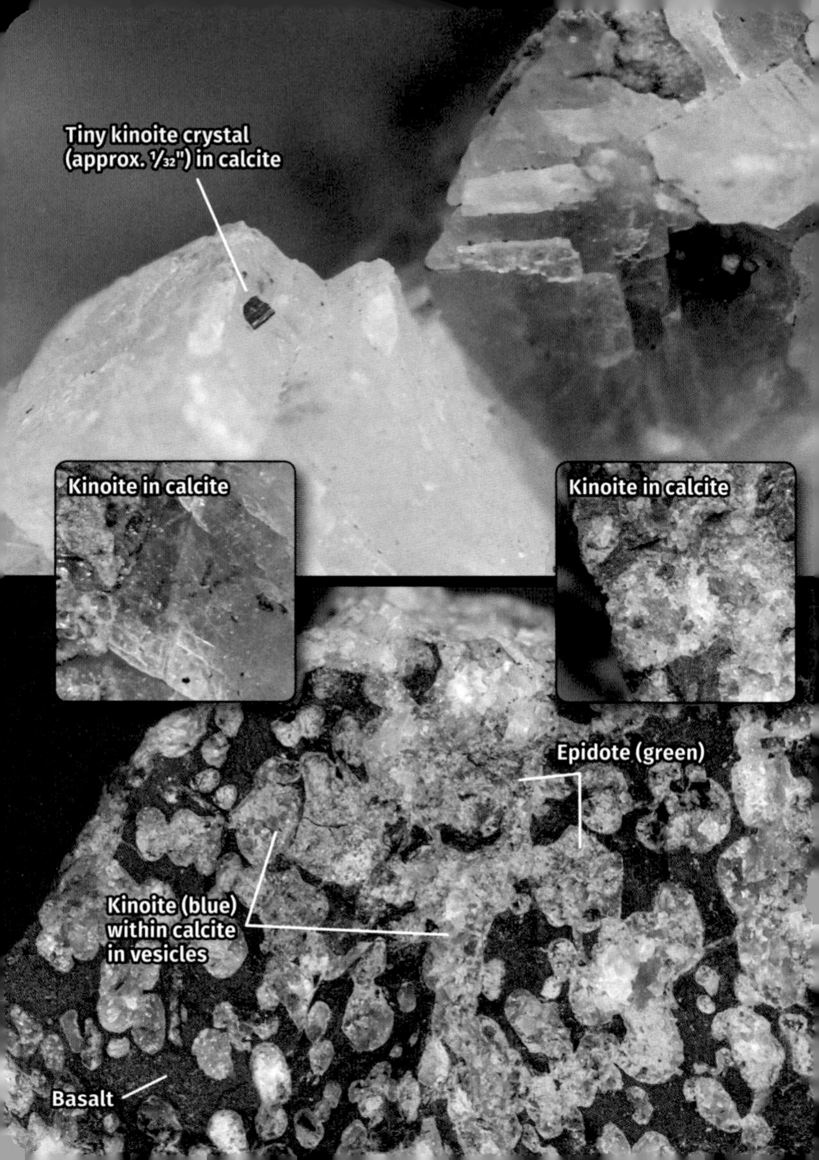

Tiny kinoite crystal (approx. 1⁄32") in calcite

Kinoite in calcite

Kinoite in calcite

Epidote (green)

Kinoite (blue) within calcite in vesicles

Basalt

Kinoite

Hardness: 5 **Streak:** Very pale blue

Primary Occurrence

Environment: Mines; rarely rivers

What to Look for: Very small, vivid blue crystals embedded in calcite or quartz

Size: Crystals are usually 1⁄16 inch or smaller

Color: Rich blue, sometimes pale blue

Occurrence: Very rare; in Michigan and Minnesota only

Notes: Kinoite is an extremely scarce copper-bearing mineral found in only a handful of places around the world—most of which are in the Keweenaw Peninsula of Michigan. Lake Superior's kinoite was formed during the same events that deposited the enormous amounts of native copper found in the region. As such, it is found in copper-rich areas, especially at old mine sites, often occurring alongside other copper-bearing minerals. But the tiny, electric-blue, elongated crystals of kinoite are rarely found free-standing and instead are almost always embedded within calcite or quartz, often visible just below the surface of those minerals as a blue smudge. This makes them even more difficult to find, but potentially more rewarding. Crystals inside quartz are stuck there permanently, but when in calcite, some carefully applied vinegar will slowly dissolve the calcite and leave the delicate kinoite crystals behind. When hunting for kinoite, look in basalt vesicles and check any pockets of quartz or calcite carefully for the characteristic blue color. Epidote and prehnite are common associates.

Where to Look: Virtually all known localities for kinoite are Keweenaw Peninsula mines; particularly famous is the Laurium Mine, 2 miles south of Calumet. The only non-Michigan locality in the region is Knife River, Minnesota, where it has been found in calcite but is very rare.

Rough limestone

Close-up of texture

Worn sample

Rough limestone

Limestone indurated (hardened) by quartz

Cavity made by fossil material

Primary Occurrence

Limestone

Hardness: 3–4 **Streak:** N/A

Environment: Rivers, shores, pits, outcrops

What to Look for: Soft, light-colored, fine-grained rock, often with a chalky feel and small hollow voids

Size: Limestone can be found in any size

Color: White to tan or brown, yellow, gray to black

Occurrence: Common

Notes: Limestone is a sedimentary rock that formed at the bottom of ancient marine seas. Millions of years ago, reefs were much larger and more common around the world, and the coral, shellfish, and microscopic organisms that made up those reefs built up into thick beds as they grew on top of each other. As the beds of organic material compressed, the calcium carbonate in the built-up shells and skeletons eventually turned into calcite that cemented it all together, becoming limestone. In the Lake Superior region, seas once collected in the depression left behind by the Midcontinent Rift and limestone formed in ample amounts. But as weathering and especially the glaciers of the past Ice Age took their toll, much of the soft rock was broken up and worn away, exposing the harder volcanic material below. Found sparingly in the region today, limestone is usually light-colored and chalky in texture and will effervesce (fizz) in acids such as vinegar, which is distinctive. It can also contain fossils, especially corals and shells. But while it is usually soft enough to be easily scratched by a knife, it can become greatly hardened by the later introduction of quartz.

Where to Look: Limestone is too soft to survive very long on shores, but is generally still widespread and can be found on beaches and adjoining rivers, particularly on the eastern half of the Upper Peninsula and in the Thunder Bay area.

Magnetite mass

Coarse, metallic surfaces

Magnetite crystal (1/8")

Magnetite mass

Magnetite grains from beach sand strongly bonding to a magnet

Primary Occurrence

Magnetite

Hardness: 5.5–6.5 **Streak:** Black

Environment: All environments

What to Look for: Metallic black grains or masses that are strongly attracted to a magnet

Size: Crystals are usually smaller than ⅛ inch; masses can be nearly any size

Color: Metallic black; rusty brown when highly weathered

Occurrence: Very common

Notes: Magnetite is an iron oxide that makes up a large component of the Lake Superior region's iron deposits, particularly in banded iron formations (page 93) and especially taconite (page 217), where it appears as tiny gray grains intermixed with chert. Masses of pure magnetite aren't common on the shore, but nearby outcrops and mines produce it; it appears as metallic masses, usually black unless weathered to a rusty brown. Such masses may appear similar to hematite (page 161), but magnetite's namesake magnetism is enough to distinguish it. In fact, magnetite is so strongly attracted to a magnet that this alone will distinguish it from almost every other Lake Superior mineral except ilmenite (page 163). Ilmenite is magnetic, too, but weakly so, and it is less common than magnetite. The region's dark volcanic rocks, especially basalt, are also rich in magnetite and may also be magnetic as a result. Upon weathering, these rocks release magnetite as dark grains of sand that concentrate on beaches.

Where to Look: Dragging a magnet through any sand or gravel, especially on the black sand beaches of Silver Bay, Minnesota, and around Thunder Bay, will yield countless samples. Some grains, under magnification, may turn out to be perfect little octahedral crystals, which is a shape resembling two pyramids placed base-to-base.

Tiny botryoidal malachite growths (green, largest approx. 1/25") within a cavity lined by prehnite crystals (pale gray-green)

Malachite

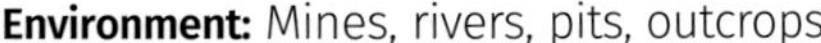

Hardness: 3.5–4 **Streak:** Light green

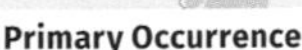

Primary Occurrence

Environment: Mines, rivers, pits, outcrops

What to Look for: Vivid green crusts or masses, sometimes with a fibrous or velvety luster, atop copper or copper-minerals

Size: Crystal clusters are usually under half an inch; crusts or masses may be up to several inches

Color: Light to dark green

Occurrence: Uncommon

Notes: Like many copper-bearing minerals, malachite forms when copper and other copper-bearing minerals are weathered and altered by water and chemicals, freeing molecules that can then recombine as new minerals. Malachite is one of the most common results of this process, and in fact any green coloration you've seen on copper roofs or coins is largely malachite. Unlike chrysocolla (page 113), with which it frequently occurs, malachite does form crystals, but they usually are very small and crude. They appear as tiny needles, often bundled together in fibrous, velvety masses. More often, it'll appear as nondescript green crusts or as little botryoidal (lumpy, grape-like) growths. Whether or not crystals or other distinct structures are present, its vivid green color and association on or near native copper (page 121) and copper-bearing minerals, especially chalcopyrite (page 107), are key for identification. Its hardness will further help you distinguish it from chrysocolla if a specimen is solid enough to be tested, though many malachite crusts are chalky and crumbly once they've been exposed to the weather.

Where to Look: You'll find malachite in any copper-bearing region. That makes the Keweenaw Peninsula mines and outcrops the most lucrative area. The Knife River area in Minnesota, and Batchawana Bay, in Ontario, are good options too.

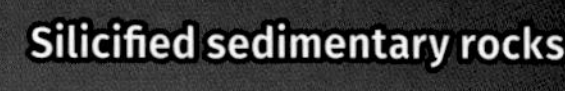

Distinct layering

Silicified layers

Water-worn slate (a metasedimentary rock derived from shale)

Specimens courtesy of Bob Wright

Metasedimentary rocks

Hardness: N/A **Streak:** N/A

Primary Occurrence

Environment: Shores, rivers, outcrops, pits

What to Look for: Rocks exhibiting features of sedimentary rocks, such as layering and coarse grains, but harder

Size: As rocks, they can be found in most any size

Color: Varies greatly; gray to black, yellow to brown, reddish brown, sometimes yellow to greenish

Occurrence: Common to uncommon

Notes: Sometimes a rock will exhibit traits that you're just not expecting it to, making identification particularly tricky. This happens most frequently with some sedimentary rocks, which in some cases may appear as you think they "should" but are notably harder. Sometimes this is the result of a sedimentary rock that has been partially metamorphosed, not quite fully transformed into a readily identifiable metamorphic rock but still exhibiting some new traits. More often, however, these "problematic" sedimentary rocks have been indurated, or hardened, by quartz. When groundwater bearing dissolved silica (quartz compound) seeps into sedimentary rocks, the silica can crystallize between the rock's grains, increasing its hardness and giving it more quartz-like properties. This is called silicification and commonly occurs in sedimentary rocks like siltstone, mudstone, and sandstone. Any fully or partially altered (changed) sedimentary rock is called a metasedimentary rock and is the most basic label you can give to any such unidentified rock. It will usually take strong magnification and additional research to pinpoint an exact rock species for your finds.

Where to Look: These kinds of rocks turn up sparingly all over the region. Water-worn slate pebbles can be found on northeastern Wisconsin's shores, and indurated siltstone and quartzite are common all along eastern Michigan and north to the Batchawana Bay, Ontario, area.

Pink margarite crystals
Chlorite vein
Mica (shiny black) in granite
Close-up of mica schist
Chlorite (dark green)
Adularia (orange)
Celadonite (bluish green) in vesicles

Primary Occurrence

Mica group

Hardness: 2.5–4 **Streak:** Colorless

Environment: Shores, rivers, pits, outcrops

What to Look for: Very soft minerals that occur as thin, flaky, flexible sheets, often so lustrous that they appear metallic

Size: Individual crystals are rarely larger than your thumbnail; usually much smaller

Color: Commonly colorless to brown, gray to black, also bluish green; very rarely pink

Occurrence: Most are common; margarite is very rare

Notes: The term "mica" is common in any book about rocks, but refers to an entire group of closely related minerals, several of which are present in the Lake Superior region, including muscovite, biotite, celadonite, and the very rare margarite. Micas are common "rock-building" minerals, meaning that they are most abundant as constituent minerals in rocks as small embedded grains. As such, they'll be easy to overlook in many cases. Mica minerals form crystals that are very thin—thinner than paper—and grow together in parallel stacks called books. Individual crystals are also very flexible and can be peeled off a book, and most micas are brightly lustrous, sometimes so "flashy" that they can appear nearly metallic. The primary way you'll find micas are as shiny dark spots in granitoids (page 157) or as "glittery" flecks in metamorphic rocks like schist (page 209). Celadonite is a blue-green mica that forms inside vesicles (gas bubbles) as a coating or lining of tiny crystals, usually with a clay-like look.

Where to Look: Micas turn up most everywhere; Minnesota and Wisconsin's shores, as well as Michigan's Keweenaw Peninsula, are all home to both mica-bearing granitoid cobbles and celadonite-lined vesicles in basalt. In Ontario, nearly any southern beach will yield granitoids with mica flakes.

Water-worn mudstone and siltstone
Soft, chalky exteriors
Right inset specimen courtesy of Bob Wright
Beach-worn siltstone
Chert layers (black)
Banded mudstone
Calcite
Siltstone layers (brown)
Siltstone

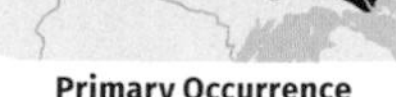

Mudstone & Siltstone

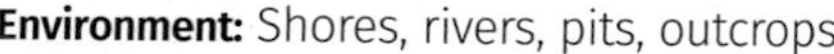

Hardness: N/A **Streak:** N/A

Primary Occurrence

Environment: Shores, rivers, pits, outcrops

What to Look for: Extremely fine-grained rocks, sometimes with a chalky or gritty feel, and may show layering

Size: They can be found in any size, from pebbles to boulders

Color: White to gray or tan, rust-yellow to brown, reddish brown to pink

Occurrence: Common

Notes: Mudstone and siltstone are two common but frequently misidentified rocks in the Lake Superior region. Like sandstone, these are sedimentary rocks named for the size of their constituent grains; siltstone contains silt-size particles measuring around 1/5,000 inch, and mudstone contains clay-size particles measuring around 1/12,500 inch. Obviously, at such a minute scale, you won't be able to see individual grains in either rock without a powerful microscope, but both rocks do tend to have a gritty, powdery, or chalky feel when freshly broken and can become exceptionally smooth when weathered. Telling them apart isn't easy, and that generally won't be necessary for novices, but distinguishing them from other kinds of rocks is easier. Both are usually fairly soft, easily scratched with a knife, though examples that have been indurated (hardened) by quartz or partially metamorphosed will be harder. Some may have colored bands caused by chemical reactions in the rock, but they do not separate along those bands like shale (page 211). Rusty siltstone can be found layered with black chert (page 109), often recessed between the weather-resistant chert layers.

Where to Look: Large formations of sedimentary rocks predominate in northern Wisconsin and Michigan's Upper Peninsula, making Lake Superior's southern shores most lucrative.

Peridotite
Green-yellow olivine grains
Olivine grains from river sand
Olivine grain in gabbro
Degraded olivine grains (rusty yellow) in highly weathered gabbro

Olivine group

Hardness: 6.5–7 **Streak:** Colorless

Primary Occurrence

Environment: All environments

What to Look for: Very hard, green or yellow masses and grains embedded in rocks like gabbro or found loose in sand

Size: Most grains or masses are smaller than a pea

Color: Yellow to green, olive green, dark gray-green to brown

Occurrence: Common

Notes: The olivine group encompasses several closely related minerals, but the two most abundant—forsterite and fayalite—are usually so indistinguishable outside of a lab that most specimens are simply labeled "olivine." Olivine is a common "rock-building" mineral, meaning that it is most abundant as a constituent mineral in rocks, especially dark rocks like gabbro. As such, it is most commonly encountered as small glassy grains or masses, usually translucent and yellow-green to green-brown in color. Embedded elongated crystals can also occasionally be found, but will be very small, and very patient searching can also reveal tiny gemmy grains in river sand, weathered from their host rock. Some feldspars (page 139) may appear similar, but olivine is slightly harder, plus its distinctive colors and occurrence in dark rocks is helpful. When highly weathered, olivine degrades, turning into opaque mixtures of other minerals, including goethite (page 155), giving rocks the appearance of "rusty" spots. Lastly, a type of rock called peridotite can occasionally be found and consists primarily of olivine, appearing similar to gabbro but with far more numerous yellow-green grains.

Where to Look: Northeastern Minnesota is home to the Duluth Gabbro Complex, one of the largest gabbro formations in the world. This makes the whole area rich with olivine-bearing gabbro cobbles and sands.

Water-worn porphyritic diabase
Feldspars (white)
Finer grained rock mass (dark)
Granite porphyry
Feldspars (orange)
Close-up of texture
Feldspars (tan)
Porphyritic rhyolite
Porphyritic basalt

Primary Occurrence

Porphyry

Hardness: N/A **Streak:** N/A

Environment: Shores, rivers, outcrops, pits

What to Look for: Rocks with large, conspicuous, angular crystals embedded within a finer grained mass

Size: As a rock, porphyry can occur in any size

Color: Usually multicolored; mottled white to gray or black, brown to reddish or purplish, greenish to yellow or tan

Occurrence: Common

Notes: As you walk Lake Superior's shoreline, porphyry is one of the few common rocks that will make you stop for a second look. But the name "porphyry" is a descriptive term that refers to a particular appearance and texture igneous rocks may exhibit, rather than to any one specific type of rock. When the magma (molten rock) that forms an igneous rock begins to cool, certain minerals crystallize and harden first, usually feldspars (page 139). If the molten rock is then thrust upward, or even erupted onto the earth's surface, the rest of the rock will cool and crystallize at a faster rate and thus produce smaller grains. The result is a rock with large, angular crystals of feldspars locked in a finer-grained mass. Various porphyritic rocks can be found; porphyritic rhyolite is abundant, usually appearing as a reddish rock with large embedded tan feldspars and glassy quartz grains. Porphyritic basalt is most common, showing clusters of slender orange crystals, usually somewhat parallel in arrangement but with some criss-crossing others. Porphyritic diabase shows large, blocky white feldspars jumbled together almost randomly, and porphyritic granite contains rectangular feldspar blocks.

Where to Look: Porphyry is abundant everywhere and most shores will yield water-worn cobbles. Minnesota's and Ontario's shores will have a particularly high concentration of finds.

Prehnite-filled cavities
in altered basalt
Iron-stained
prehnite mass
"Patricianite"
Water-worn
prehnite masses
Vein of prehnite removed
from host rock
Fine crystals
Rough, iron-stained
prehnite growth
Water-worn crystal clusters

Prehnite

Hardness: 6–6.5 **Streak:** White

Primary Occurrence

Environment: Shores, rivers, outcrops, pits

What to Look for: Hard, pale green masses or rounded clusters of crystals coating or embedded within rocks

Size: Masses and crusts can range from inches to feet in size

Color: Commonly pale green; also white to gray, brownish, pink

Occurrence: Uncommon to common

Notes: Prehnite is a popular and fairly abundant collectible mineral. It can form several ways, but in the Lake Superior region it was typically a result of the same hydrothermal (hot, rising, mineral-rich water) activity that produced many of the region's other collectibles, such as copper and agates. This makes it a relatively common vesicle- (gas bubble) and vein-filling mineral within volcanic rocks, particularly basalt (page 97). When finely formed, it develops small wedge-shaped crystals, usually arranged into radial, ball-like groups that give crusts and masses of prehnite a lumpy appearance. When broken, these growths exhibit an almost fibrous, fan-shaped cross-section. Massive growths that fill a crack or cavity completely, often within purplish basalt, are harder to identify, but can also exhibit the same cross-section texture. In most cases, its color will be distinctive; prehnite is typically pale green in color, and usually glassy and translucent. Many specimens occur with calcite and zeolites, and may be stained brown by iron. It can resemble quartz (page 197), but prehnite is softer. Lastly, a copper-bearing variety known locally as "patricianite" is found in Michigan and exhibits pale pink to reddish coloration mottled within the typical shades of green.

Where to Look: Look in basalt formations; Minnesota's shores and adjacent rivers from Duluth to Silver Bay are lucrative. "Patricianite" is found in Michigan's Keweenaw Peninsula, on the Copper Harbor and Fort Wilkins-area beaches.

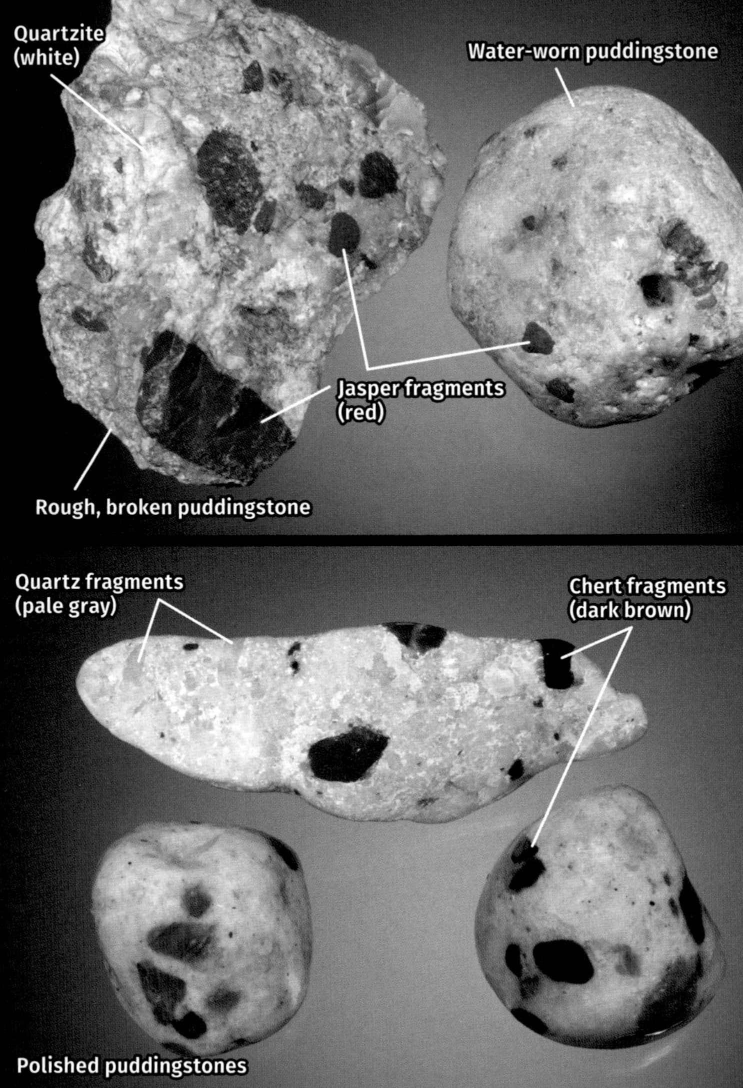
Quartzite
(white)
Water-worn puddingstone
Jasper fragments
(red)
Rough, broken puddingstone
Quartz fragments
(pale gray)
Chert fragments
(dark brown)
Polished puddingstones

“Puddingstone”

Hardness: ~7 **Streak:** N/A

Primary Occurrence

Environment: Shores, rivers, pits

What to Look for: Very hard, white rock containing embedded pieces of bright red and black material

Size: Typical specimens are smaller than a softball

Color: Multicolored; white to gray or tan base color, with red to brown and black spots

Occurrence: Uncommon; only in Michigan and Ontario

Notes: Undoubtedly one of eastern Lake Superior’s most beloved rocks, puddingstone has long been popular as a souvenir of the area’s beaches. This hard, predominantly white rock began as loose sediments, primarily quartz-rich sand but also gravel-size pebbles of red jasper (page 165), black-brown chert (page 109), and white quartzite (page 203). These sediments did not settle into neat layers but were instead jumbled together in a river setting. Over time, this material hardened to become conglomerate (page 119), a sedimentary rock. With later heat and pressure, the conglomerate was then partially metamorphosed, fusing the sand grains and locking the whole mass together into the quartzite-rich metasedimentary rock (page 177) that we call “puddingstone.” It gets its peculiar name from early settlers who thought it resembled their bread pudding; the characteristic red and black pebbles in it do look a bit like fruit and nuts. But Lake Superior’s puddingstone didn’t form here; it originated several miles east and southeast, in Ontario, and was picked up and deposited by the glaciers.

Where to Look: In the Lake Superior region, only the far southeastern shores yield puddingstone, primarily in Chippewa County, Michigan, and the Sault Ste. Marie, Ontario, area, as well as along St. Mary’s River and southward.

Whole nodule in basalt
Water-worn
nodule in basalt
Polished
nodules
Water-worn nodule
in basalt
Water-worn
"greenstone" nodules
Epidote (dark
yellow-green)
Chlorastrolite
(polished)
Tiny pumpellyite crystals
(gray-green)

Pumpellyite & “Greenstone”

Hardness: 5.5–6 **Streak:** White

Primary Occurrence

Environment: Shores, rivers, mines, pits

What to Look for: Small green clusters of needle-like crystals; dark green nodules or pebbles with a spiderweb pattern

Size: Growths of pumpellyite are rarely larger than a few inches

Color: Yellow-green or olive green, bluish green to dark green, gray-green, occasionally almost black

Occurrence: Uncommon in Michigan; rare elsewhere

Notes: Pumpellyite is a popular collectible that forms in cavities in basalt and diabase as the rock is affected by hot mineral-bearing groundwater. Fine crystals can be found in the region and appear as tiny needles, usually arranged into radial clusters and intergrown in complex crusts. Crystals are typically found in basalt cavities alongside quartz and epidote (page 137), a mineral to which it is closely related. The delicate nature and typical gray-green coloration of crystals will distinguish it from the more robust and typically yellow-green crystals of epidote. But pumpellyite is also frequently found in nodular (round clusters) form, which formed when tightly intergrown, nearly microscopic crystals filled vesicles (gas bubbles) and cavities completely. Called “greenstone,” these masses can be found still embedded in their host rock or worn free as thumbnail-size beach pebbles and can be confused with chlorite (page 111), but chlorite is much softer. A nodular variety known as chlorastrolite features beautiful spiderweb or “turtle-back” patterns and is a prized local gemstone.

Where to Look: The basalts of the Keweenaw Peninsula are the primary source for both crystals and nodules. The mine dumps in the Osceola and Phoenix areas are especially lucrative. Isle Royale was once famous for chlorastrolite nodules, but the island is a National Park and collecting there is illegal.

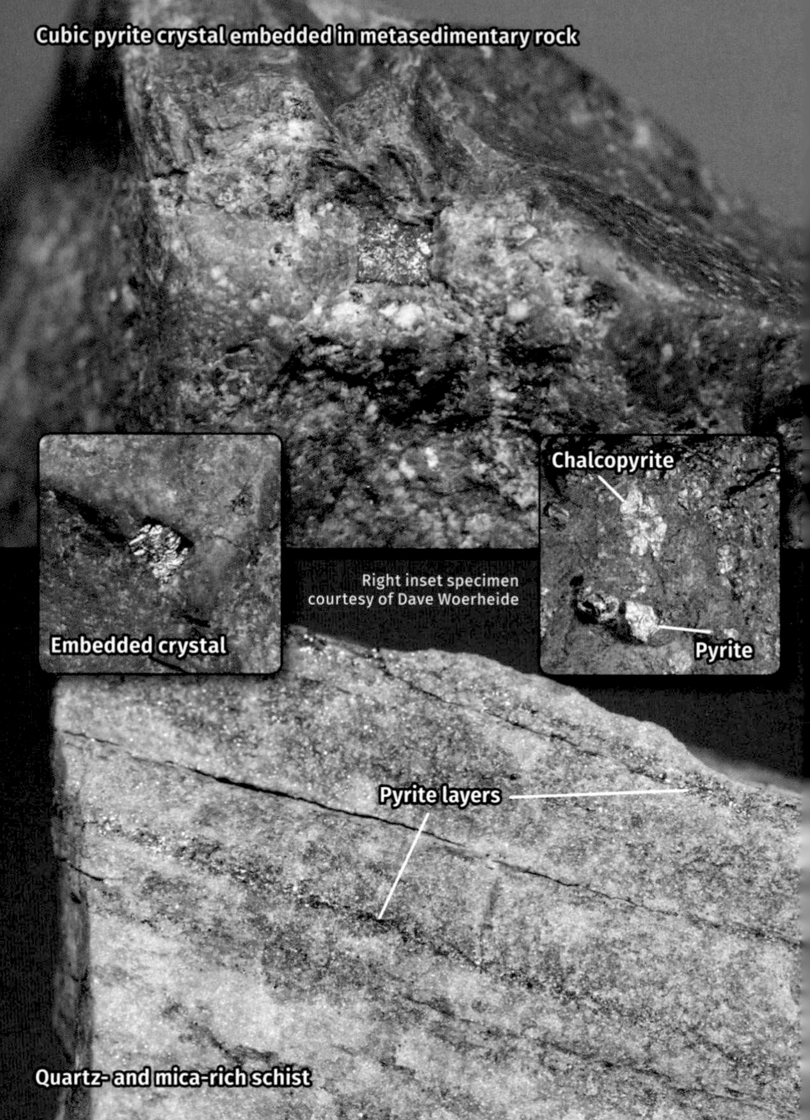
Cubic pyrite crystal embedded in metasedimentary rock
Embedded crystal
Chalcopyrite
Pyrite
Right inset specimen courtesy of Dave Woerheide
Pyrite layers
Quartz- and mica-rich schist

Primary Occurrence

Pyrite

Hardness: 6–6.5 **Streak:** Green-gray to black

Environment: All environments

What to Look for: Hard, brassy, metallic yellow masses, grains, or cubic crystals embedded in layered rocks

Size: Crystals are rarely larger than your thumbnail

Color: Pale brassy yellow, sometimes rusty brown

Occurrence: Uncommon

Notes: As one of the most widespread iron-bearing minerals, pyrite can be found in many different kinds of geological environments, especially in sedimentary and metamorphic hosts. While fine crystals are not something you'll pick up on a beach, it does turn up as small but conspicuous embedded masses, so learning to recognize its traits will be helpful. It is almost always a pale yellow brassy color with bright metallic luster unless weathered, in which case it develops a dull, rust-colored coating that can also stain nearby rock. When crystallized, pyrite forms cubic shapes, often with striated (grooved) sides. Free-standing crystals are not common around Lake Superior, but you may find them embedded in rocks where they'll appear as metallic squares. Irregular grains or small masses are the typical way you'll find pyrite, especially within cavities in sedimentary rocks like limestone (page 171), and as "sparkly" layers in metamorphic rocks like schist (page 209). Chalcopyrite (page 107) is very similar in appearance, but is much softer and doesn't form as cubes. Pyrite's longtime nickname has been "fool's gold," but its relative abundance, pale color, high hardness, and brittle nature mean you'll never actually confuse it for gold.

Where to Look: Pyrite is not a typical shore find, though it can be found in some metasedimentary rocks along Minnesota and Wisconsin's shores. Michigan's far eastern shores may produce better specimens from sedimentary rock outcrops, and in Ontario, schist formations from Terrace Bay to Marathon, just inland from Lake Superior, produce pyrite.

Blocky pyroxenes (black) in gabbro
Parallel, step-like breaks
Augite crystal in cut gabbro
Augite crystals (black) in gabbro
Augite crystals (black) in gabbro

Pyroxene group

Hardness: 5–6 **Streak:** Greenish gray to white

Primary Occurrence

Environment: Shores, rivers, outcrops, pits

What to Look for: Dark-colored, blocky masses or grains embedded in rocks, especially dark coarse-grained rocks

Size: Most specimens are thumbnail-size or smaller

Color: Black to brown, greenish gray to dark green

Occurrence: Common

Notes: Much like their mineral cousins, the amphibole group (page 89), the pyroxene minerals are a family of rock builders, appearing most often as constituent minerals in igneous rocks, especially gabbro. There are many pyroxenes, but only a few are prevalent in the Lake Superior region, including augite, aegerine, and diopside, although if fine crystal shapes are not present, it will be very difficult to distinguish them from each other outside a lab. Augite and aegerine are perhaps the most common, usually appearing as black glassy grains embedded in rocks like gabbro and many granitoids. Ideally, these minerals have a blocky crystal shape something like a "leaning" rectangle (or, more rarely, as elongated needles), but this is not always visible. When indistinct, they are easily confused with amphibole minerals, but there are a few key traits to look for. Pyroxenes tend to have a glassier luster, as opposed to the typical fibrous texture of many amphiboles. They also have cleavage at nearly 90-degree angles, meaning that when pyroxenes break, they do so in perpendicular, blocky, step-like patterns, while amphiboles break in sharper, more jagged angles.

Where to Look: The Duluth Gabbro Complex, comprising much of northeastern Minnesota and with exposures near Lake Superior, is the prime spot to look for embedded pyroxene crystals. But they are ubiquitous in rocks all over the region.

Quartz crystal cluster (largest approx. ½")

Druzy quartz

Left inset specimen courtesy of Jim Cordes

Quartz-lined basalt vesicle

Curved cracks

Quartz (gray-white) in granitoid

Water-worn quartz

Druzy quartz on limestone

Quartz

Hardness: 7 **Streak:** White

Primary Occurrence

Environment: All environments

What to Look for: Abundant, light-colored, very hard translucent masses, veins, or six-sided, pointed crystals

Size: Crystals can be up to a few inches; masses can be any size

Color: Colorless to white or gray; commonly yellow to brown or red when impure, rarely greenish

Occurrence: Very common

Notes: Making up 12 percent of the earth's crust and present in a majority of rock types, quartz is rivaled only by feldspars as the most abundant mineral on the planet, and it is perhaps the most important mineral for collectors to be able to identify. It consists purely of silica—a combination of silicon and oxygen that is a common ingredient in many minerals—and as such is very hard, glassy, brittle, has a conchoidal fracture (when struck, the shape of the crack or break is curved or circular), and it is usually colorless to white unless stained by other minerals. Crystals are common and appear as six-sided, elongated prisms with pointed or wedge-shaped tips, sometimes formed singly but more often in groups or crusts of tiny crystal points called druse. You'll find crystals within cavities in a wide variety of rocks, especially vesicles (gas bubbles) in basalt, fissures in rhyolite, and pits in limestone. But quartz is more common as masses and veins filling cavities, and it is most common of all as constituent grains in rocks, especially granite. Quartz could be confused with calcite, but is much harder, more abundant, and won't effervesce (fizz) in acids such as vinegar.

Where to Look: Mine dumps in the Keweenaw of Michigan will produce fine little crystals in basalt cavities. Sedimentary rocks in northern Wisconsin and eastern Michigan will produce druse, and Ontario granitoids are full of quartz.

Quartz vein in basalt

Quartz vein in fragmented rock

Quartz druse in vesicle

Quartz crusts

Quartz-filled vesicle in basalt

Water-worn quartz

Waxy luster

Carnelian pebbles from lakeshore (largest approx. ½")

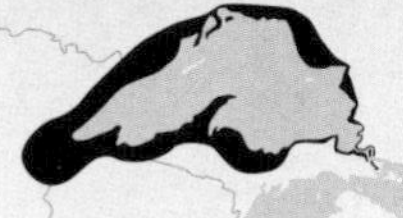

Primary Occurrence

Quartz, varieties

Hardness: 7 **Streak:** White

Environment: All environments

What to Look for: Abundant, light-colored, very hard translucent masses, veins, or six-sided, pointed crystals

Size: Crystals can be up to a few inches; masses can be any size

Color: Colorless to white or gray; commonly yellow to brown or red when impure, rarely greenish

Occurrence: Very common

Notes: The abundance of quartz and its presence in nearly every kind of rock environment means that it also has many variations in structure and means of occurrence. Since it is commonly deposited by hydrothermal activity (hot, mineral-rich groundwater rising into overlying rock), quartz can make its way into the cavities and pores within many different materials. You'll frequently find it as vein fillings in igneous rocks and as the "glue" that cements grains of sediment together in some sedimentary rocks. Quartz is also particularly common in its various microcrystalline forms. These varieties of quartz don't have an outwardly visible crystal shape but are instead comprised of countless tiny, microscopic grains (as in jasper, page 165) or fibers (as in chalcedony, page 103) grown into hard masses with waxy luster. Microcrystalline forms often fill their host vesicles or other pockets completely, taking the shape of their surroundings. The most collectible microcrystalline quartz varieties are agate (page 39) and carnelian, or red chalcedony. Lastly, quartz is also very common as rounded white, translucent beach pebbles, weathered into rounded shapes after glaciers and other erosive forces freed masses from their host rock.

Where to Look: Veins, masses, and microcrystalline varieties like chalcedony are ubiquitous and found everywhere with little effort. The many lava flows on Minnesota's shores make for countless cavities and cracks where quartz could form.

Hematite-coated amethyst crystal clusters
Richly colored crystal
Fine crystal cluster
with red hematite
Smoky quartz cluster
Pale water-worn
mass
Chalcopyrite
Amethyst
mass
Color zoning
Rough, broken fragments

Quartz, amethyst

Hardness: 7 **Streak:** White

Primary Occurrence

Environment: All environments

What to Look for: Very hard, glassy, translucent, purple pointed crystals, often clustered together, or masses

Size: Masses of amethyst can grow up to several feet

Color: Light to dark purple, often with white sections; frequently with red-brown surface coloration; also dark gray

Occurrence: Uncommon in Ontario; generally rare elsewhere

Notes: Amethyst is among the most popular varieties of quartz, and the Lake Superior region is particularly rich with fine specimens. Unlike some colorations seen in quartz that are caused by mineral impurities staining the crystals, amethyst's color is derived from within, caused by naturally irradiated iron atoms trapped within the microscopic spaces in quartz's crystal structure. The result is the trademark purple color that makes amethyst unique, often with internal white or gray "smoky quartz" zones caused by periods of varying amounts of impurities during crystal growth. (Smoky quartz, generally rare in the region, is similar but gets its gray-black color from irradiated aluminum). Amethyst can be found in agates and occasionally as pale water-worn pebbles, but the most impressive crystals are found in Ontario. The famous mines of the Thunder Bay area dig into brecciated granitoid (page 157) masses to find sometimes enormous pockets and veins lined with finely formed amethyst crystals, usually infilled with clay. These crystals frequently have red hematite spots on their surfaces, sometimes coating entire crystal clusters.

Where to Look: The chief region for Lake Superior amethyst is the Thunder Bay, Ontario, district, especially around Pearl, where several pay-to-dig mines operate, and eastward toward Rossport. Elsewhere, it can be found sparingly as water-worn pebbles or filling in the centers of agates.

Water-worn quartzite
Close-up of texture
Flaky breaks
Iron-stained quartzite
"Puddingstone" (quartzite conglomerate)

Quartzite

Hardness: ~7 **Streak:** N/A

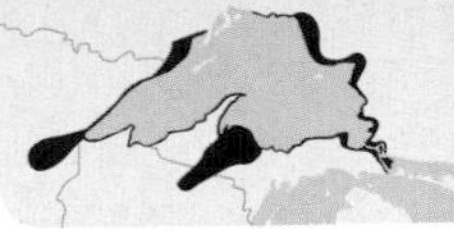

Primary Occurrence

Environment: Shores, rivers, pits, outcrops

What to Look for: Very hard light-colored rock with a generally even color and texture, sharing many of quartz's traits

Size: As a rock, quartzite can be found from pebbles to boulders

Color: White to gray, yellow to brown, pink to orange

Occurrence: Common

Notes: Quartzite is a metamorphic rock derived from sandstone (page 207). Sandstone is composed of compacted sand, and sand is predominantly made up of grains of quartz. When sandstone underwent compression and heating, the quartz grains fused together into a solid, hard mass. In other cases, silica solutions carried by groundwater seeped into the sandstone, crystallizing as quartz and locking the sand grains together. Through either process, the result is quartzite, a metasedimentary rock so rich with quartz that it shares many of its traits, namely its high hardness and weather resistance. It is usually white or gray, but is often stained yellow or reddish with iron minerals. You could confuse it with chert (page 109) or quartz (page 197) itself, but its granular structure is distinctive. On close inspection, you'll be able to see the tiny glassy quartz grains that make up quartzite, which give it a translucency that will distinguish it from opaque, waxy chert, and a grainy surface texture and flaky chipping that will distinguish it from pure quartz. In addition, quartz will be smooth and glassy on a fresh break while quartzite will have rough, grittier surfaces when broken.

Where to Look: Quartzite is found as water-worn cobbles all over the region. Eastern Michigan's shores are home to puddingstone (page 189) pebbles, composed largely of quartzite, and the areas around Michipicoten Bay in Ontario and Marquette in Michigan have sizable quartzite outcrops.

each-worn rhyolite

Parallel flow banding

Vesicles

Close-up of texture

Vesicular rhyolite

Parallel flow banding

Mineral-filled vesicles

ineral-filled vein

Beach-worn rhyolite

Primary Occurrence

Rhyolite

Hardness: 6–6.5 **Streak:** N/A

Environment: Shores, rivers, pits, outcrops

What to Look for: Fine-grained, reddish gray rock, often with subtle horizontal bands and/or vesicle (gas bubbles)

Size: Rhyolite can be found in any size, from pebbles to cliffs

Color: Light to dark gray, often reddish brown to red-gray, also brown to tan, sometimes with subtly colored banding

Occurrence: Very common

Notes: Rhyolite is a fine-grained volcanic rock that formed when molten rock was erupted onto the earth's surface where it could cool rapidly. It contains the same mixture of minerals found in granite (page 157), but granite cooled slowly, deep within the earth, which allowed its minerals to grow to a large, visible size. While rhyolite may initially appear quite similar to basalt (page 97), rhyolite's lava (molten rock) was thicker and stickier than that which formed basalt, resulting in some differences that will help distinguish it. Even though it cooled quickly, its thick lava held heat longer than basalt's, so despite still being very fine-grained, rhyolite does exhibit some visible grains and embedded crystals under magnification. The thick, slow-flowing lava also produced layers in some rhyolite formations, appearing as wavy but generally parallel bands. Also like basalt, rhyolite frequently exhibits many vesicles, bubbles formed by trapped gases, which may be filled with calcite, quartz, or sometimes baryte, as well as chalcedony or agate in some areas. It could be superficially confused with mudstone (page 181), but it has larger grains.

Where to Look: Rhyolite is found in many places around Lake Superior, frequently as pebbles on beaches. Enormous rhyolite formations can be seen on Minnesota's shores; Palisade Head and Shovel Point are two famous rhyolite flows.

Right inset specimen courtesy of Jim Cordes

Sandstone

Hardness: N/A **Streak:** N/A

Primary Occurrence

Environment: Outcrops, pits, rivers, shores

What to Look for: Rough, gritty rocks that feel as if they are made of sand, often with layering and variegated color

Size: Sandstone can be found in any size, from pebbles to cliffs

Color: White to gray, tan to brown, pink to reddish orange

Occurrence: Very common

Notes: Like many sedimentary rocks, sandstone is named for the particles that compose it. When sand (tiny mineral fragments weathered free from rocks) was transported by rivers and deposited in large quantities in seas or lakes, it built up in thick beds. Over time, as the weight of the upper beds compressed the lower ones, the sand became tightly compacted. Other minerals, such as calcite and clays, were deposited by groundwater and crystallized between the grains of sand, cementing it all into one hard mass. The result is sandstone, one of the easiest rocks to identify. Because sand is composed largely of quartz, sandstone has a gritty sandpaper-like feel, even when weathered. It also has a distinctly grainy appearance, and individual grains of sand are often easily rubbed or picked free from the stone. Colored layers are common in sandstone and formed by repeated deposition of sand; colored rings or spots are common in limestone as well, but formed by chemical reaction after the rock had formed. There isn't much that you could confuse for sandstone; graywacke (page 159) shares many traits, but it is generally more fine-grained and darker in color, and quartzite (page 203) may look similar at first glance, but it is much harder, more solid, and may be translucent.

Where to Look: Northern Wisconsin and Michigan's Keweenaw Peninsula are largely underlain by sandstone; both water-worn cobbles and large outcroppings are very common.

Beach-worn mica schist

River-worn mica schists

Flaky layered edge

Rougher layered edge

Close-up of texture

Pyrite layers

Rough, broken piece of quartz-mica schist showing parallel layers

Schist

Hardness: N/A **Streak:** N/A

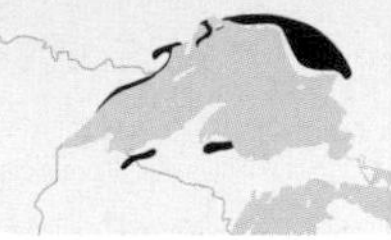

Primary Occurrence

Environment: All environments

What to Look for: Generally fine-grained rocks with tight, thin layering, often full of thin, shiny mineral flakes

Size: Schist occurs in any size, from pebbles to boulders

Color: Usually multicolored; gray to black, white, tan to brown, greenish gray, sometimes pinkish

Occurrence: Common

Notes: Like gneiss (page 153), schist is a metamorphic rock that forms when older rocks are subjected to heat and pressure within the earth. Schists can develop from a variety of parent rocks, but most commonly form when sedimentary rocks like shale and mudstone are metamorphosed. As the parent rock is buried and heated, the forces acting upon it soften its mineral grains, even changing them into new minerals, while pressing them into distinct layers. This is similar to how gneiss forms, but where schist differs is that more than half of its minerals have been turned into flaky, plate-like minerals (most often micas, but also chlorite and amphiboles) and arranged into tightly packed parallel layers. This gives schist a more layered, flaky appearance and is often loaded with "glittery" mineral grains. Schist can also contain collectible minerals formed during metamorphosis, such as garnets and pyrite. Schist's parallel structure and flaky grains mean it is usually easy to identify, but water-worn pieces are trickier; many appear smooth with "smears" of color, but they will still show hints of layering or flakiness on their edges.

Where to Look: Schist is common; many specimens are glacially deposited. In Ontario, you can find formations from Terrace Bay to Marathon, just inland from the lake. In Michigan, the Ishpeming and Marquette areas are rich with schists.

Shale
Thin, flat
separated layers
Close-up of texture
Edge of one
single layer
Shale

Shale

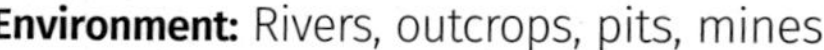

Hardness: >5.5 **Streak:** N/A

Primary Occurrence

Environment: Rivers, outcrops, pits, mines

What to Look for: Fairly soft, fine-grained, with many flaky layers that can be separated with a knife

Size: As a rock, shale can be found in any size

Color: Tan to brown, yellow-brown, gray to black

Occurrence: Common

Notes: Shale is a common sedimentary rock that formed when very fine grains of weathered rocks and minerals settled to the bottom of very calm bodies of water, usually lakes. Over time, repeated (often seasonal) periods of sedimentation created distinct layers, and later compaction by the weight of overlying sediments solidified the layers together to form shale. The individual grains that make up shale are generally clay-size and are far too small to see without serious magnification, but in some specimens you may be able to spot the telltale "glitter" of countless tiny fakes of mica minerals (page 179). The fine-grained nature makes shale fairly soft, easily scratched with a knife, and gives it a gritty, dirty feel on fresh breaks. But its layers are its defining characteristic. Shale will split and break along its layers, and most specimens will be thin plates. By using a knife blade you can easily separate the layers of larger masses. Similarly super fine-grained sedimentary rocks lacking distinct layering are by definition not shale—they are mudstone or siltstone (page 181). Lastly, when subjected to heat and pressure, shale becomes slate (see metasedimentary rocks, page 177), which may have a similar appearance to shale but is harder, more tightly layered, and usually more solid.

Where to Look: The Porcupine Mountain, Michigan, area is home to some impressive shale exposures along area rivers. Shale also turns up at exposures all across the region, especially the Keweenaw Peninsula and eastern Michigan shores.

Rough silver
Sharp, gnarly edges
Acanthite surface coating (gray)
Dendrites in quartz
Silver mass on basalt
Silver dendrites (coated in gray acanthite) in beach-worn quartz

Silver

Hardness: 2.5–3 **Streak:** Metallic silvery gray

Primary Occurrence

Environment: Mines, rivers, outcrops, pits

What to Look for: Soft, white silvery metal, usually with a dull, dark gray surface coating; often with copper

Size: Most silver specimens are no larger than your palm

Color: Silvery white to gray when freshly exposed; dark gray to brownish exteriors when weathered

Occurrence: Rare in Michigan and Ontario; very rare elsewhere

Notes: Like copper (page 121), silver is a native element, consisting not of a mixture of elements, as in most minerals, but solely of silver atoms. This famously valuable metal was deposited in cavities and veins within the region's rocks during the same hydrothermal (hot, mineral-rich groundwater and steam) events that formed the region's copper. As such, silver is most often found as irregular masses or veins that are usually rough, gnarly, and sharp-edged; it is often found in quartz but also within basalt and other rocks. Free-standing crystals are present in the region but are extremely rare; most will appear as blocky, coarse, tree-like growths. Little branching crystals called dendrites may rarely be found as well, and they occur embedded in rock or quartz. But in all cases, silver is best identified by its low hardness, malleability, and, of course, its trademark metallic silver color. Most specimens that have been exposed to air and weather will be coated in a dull, dark mineral called acanthite, better known as tarnish. Though easily scratched away to reveal the silver color below, the coating can make specimens easy to overlook.

Where to Look: Michigan's Keweenaw Peninsula copper mine dumps are the most lucrative, especially north of Houghton. The famous Silver Islet area, at the tip of Sibley Peninsula, Ontario, still occasionally produces specimens in quartz.

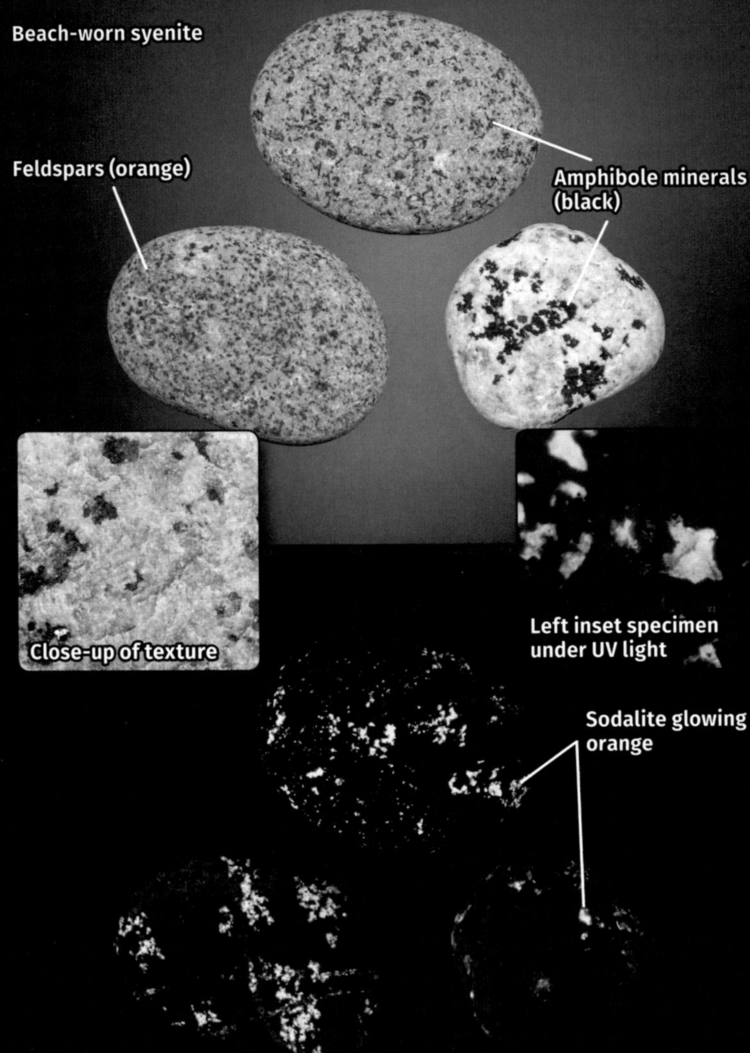
Beach-worn syenite
Feldspars (orange)
Amphibole minerals (black)
Close-up of texture
Left inset specimen under UV light
Sodalite glowing orange
Same specimens as above under shortwave UV light

Syenite

Hardness: N/A **Streak:** N/A

Primary Occurrence

Environment: Shores, outcrops

What to Look for: Light-colored, mottled rocks greatly resembling granite but partly glow orange under UV light

Size: Most water-worn cobbles are smaller than a foot

Color: Varies; mottled gray, black, white, and orange-pink

Occurrence: Rare

Notes: Syenite is a rare rock on Lake Superior's shores, and while it may look a lot like granite (page 157), it differs greatly from it and other granitoid rocks because of its distinct lack of quartz. This happened in a very particular rock-forming scenario: syenite formed from a partially solid magma in which the available silica was largely used up by the formation of potassium-feldspars instead of quartz. This may seem like a mundane fact, but this process also allowed for rare minerals such as sodalite to form within the rock. How that becomes interesting for collectors is that the otherwise gray sodalite in Lake Superior's syenite is highly fluorescent under ultraviolet (UV) light, glowing bright orange. This is syenite's primary identifying feature for collectors. Fluorescent pebbles, often collected at night by using a UV flashlight, are wildly popular on Michigan's shores where they are known locally as "Yooperlites®." Studies have shown that Michigan's pebbles originated from a large formation of Canadian syenite called the Coldwell Complex on Lake Superior's northern shore, and they were carried south by glaciers.

Where to Look: Pebbles are found along the southern shores of Lake Superior, especially the eastern half of Michigan's Upper Peninsula, where they are best hunted with a UV light. Syenite pebbles do turn up in northern Wisconsin and more rarely on Minnesota's shores, as well. The Coldwell Complex can be seen at outcrops about 20 miles west of Marathon, Ontario, along Highway 17.

Rough taconite mass
Specimen courtesy of John Woerheide
Taconite pellets
Taconite pellet cluster
Smooth surface from breaking along layers
Faint layering
ough taconite

Taconite

Hardness: 6–7 **Streak:** N/A

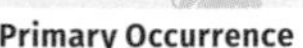

Primary Occurrence

Environment: Pits, mines, outcrops, rivers

What to Look for: Dense, hard, dark rock that is highly magnetic; sometimes with visible layering

Size: Masses can be any size, from inches to feet

Color: Dark gray to black, sometimes brown

Occurrence: Uncommon

Notes: Taconite is a variety of banded iron formation (page 93), and as such contains fine grains of iron-bearing minerals interspersed within layers of gray chert (page 109). The predominant iron mineral in taconite is magnetite (page 173), and it is mined extensively at inland iron ranges in Minnesota and in Michigan. As an ore material, it's not particularly rich—it consists of only 20-30 percent iron—but it is plentiful enough that it is a viable source of iron, and in fact is the primary source of domestic iron in the United States. Hard, dense, and dull gray taconite is easily overlooked by collectors, though its faint layering, magnetism, and faintly metallic, "glittery" grains will help identify loose finds. Large exposures may initially appear to be a dark volcanic rock, but the lack of vesicles (air bubbles) in taconite, combined with its other traits, will easily distinguish it. Near iron-ore transportation routes, you may find taconite pellets—small metallic balls the size of marbles—which are a man-made form of refined taconite combined with clay, shaped for easy transportation. But beware of trespassing on railroads.

Where to Look: The Gunflint Range of northern Minnesota and particularly in Ontario, near Thunder Bay, produce tons of taconite, and masses can be found in area outcrops and rivers. In Michigan, the iron mines near Marquette produce taconite, so the surrounding area will yield specimens.

Rough, gnarled copper mass
Tenorite surface coating (dark brown-black)
Tenorite (dark) on copper crystal
Right inset specimen courtesy of A. E. Seaman Mineral Museum
Solid mass of tenorite
Copper veins coated in dark tenorite
Conglomerate host rock

Primary Occurrence

Tenorite

Hardness: 3.5 **Streak:** Black

Environment: Mines, outcrops, pits, rivers

What to Look for: Black masses or crusts, sometimes metallic, coating or nearby copper

Size: Coatings or masses are rarely larger than a few inches

Color: Dark gray to black, sometimes very dark brown

Occurrence: Common

Notes: A simple combination of copper and oxygen, tenorite is a common product of weathered deposits of native copper (page 121). While not as colorful or collectible as malachite or chrysocolla, tenorite is similarly found atop or very closely associated with copper, developed as water and other elements combined with the native metal. As such, tenorite is virtually always found as a coating on copper, usually thin enough to be somewhat easily scratched or cleaned off to reveal the red metal below, but it can occasionally be thicker and tougher. Solid masses, up to several inches in size, are rarer but have been found at old mine sites in Michigan. Such masses may be pure enough to have a nearly metallic appearance, but most examples of tenorite are fairly dull. In general, specimens are fairly uninteresting and easily overlooked unless associated with a particularly striking formation, such as a copper crystal, or with other, more colorful copper-bearing minerals like cuprite, with which it provides excellent color contrast. Identification is fairly easy, given its low hardness, black streak, and almost invariable association with copper.

Where to Look: You'll find tenorite most places copper is found, making the old mines of the Keweenaw Peninsula your best bet. Near-shore outcrops, like the Copper Harbor Conglomerate, especially produce copper blackened by tenorite.

Unakite beach pebbles
Close-up of texture
Unakite beach pebble
Epidote (green)
Feldspar (orange)
Quartz (grayish)

Unakite

Hardness: N/A **Streak:** N/A

Primary Occurrence

Environment: Shores, rivers, pits

What to Look for: Beach pebbles resembling granite but of distinctly green and orange or pink coloration

Size: Unakite beach stones are rarely as large as your fist

Color: Multicolored; yellow-green to dark green and orange to reddish or pink, usually with some white or colorless spots

Occurrence: Uncommon

Notes: Unakite is a granitoid (page 157) uniquely identifiable by its coloration alone. Consisting primarily of tightly bonded masses of yellow-green epidote (page 137), orange or pink potassium feldspars (page 139), and white or colorless quartz (page 197), unakite is full of bright, contrasting colors that make it a popular collectible. While it shares the texture and structure of granite, it is actually an example of what is called a metasomatic rock. In metasomatism (a type of metamorphic activity), mineral-rich groundwater and other chemicals alter a preexisting rock, changing some (or all) of its minerals into different minerals without changing the texture of the original rock. In the case of unakite, many of the original granitoid's calcium- and aluminum-bearing minerals turned into green epidote, while the orange feldspar and white quartz remained unchanged. Even in very pale specimens, little else but visual inspection is needed to identify it. Lake Superior's unakite is found exclusively as water-worn pebbles, deposited in the region by the glaciers of the past Ice Ages after being carried southward from their source in northern Canada.

Where to Look: Unakite turns up sparingly on most shores, perhaps with most frequency along Ontario's and Minnesota's cobbled shorelines and adjoining rivers.

Radial zeolite crystal clusters in host rock

Lustrous, freshly exposed crystals

Dull, water-worn surfaces

Needle-like crystals

Heulandite crystal (approx. 1/16")

Zeolite-filled vesicle

Fan-shaped crystal clusters

Natrolite crystal clusters in host rock

Zeolite group

Hardness: 3.5–5.5 **Streak:** White

Primary Occurrence

Environment: Shores, rivers, outcrops

What to Look for: Small, light-colored, brittle and often fibrous or radial crystals filling cavities in dark volcanic rocks

Size: Individual crystals tend to be small, under an inch; crystal groups or masses may be up to fist-size or rarely larger

Color: Varies; typically colorless to white or gray, tan to brown, yellowish to orange, pink to salmon, green; can be banded

Occurrence: Common to rare, depending on the mineral species

Notes: The zeolite group encompasses dozens of closely related minerals, and while Lake Superior is home to a number of them, only a handful are abundant. In most cases, including the majority of occurrences around Lake Superior, zeolites form within aluminum-bearing igneous rocks, particularly basalt and diabase, when they are affected by low-temperature and low-pressure metamorphic processes. After the host rock's formation, mineral-bearing groundwater can alter it, pulling elements from the rock's constituent feldspars and creating alkaline conditions conducive to mineral growth within the vesicles (gas bubbles) and other cavities in the rock. Along with chlorite (page 111) and quartz (page 197), zeolites are a primary result of this process, and often filled the cavity completely. Many of the most recognizable zeolites, like thomsonite (page 235), develop fine, needle-like crystals, usually arranged into radial or fan-shaped "sprays" with a fibrous luster when freshly exposed. Others, like rare heulandite, form blockier, glassier crystals.

Where to Look: Zeolites will turn up almost anywhere you find basalt, making the western half of Lake Superior's shores—especially the entirety of Minnesota's shores and much of the Keweenaw Peninsula—your best bet.

Zeolite crystal cluster in weathered host basalt

Thomsonite growths

Chlorite-filled center

Thomsonite in basalt vesicle

Clusters of needle-like zeolite crystals encased within calcite mass

Specimen courtesy of Dave Woerheide

Zeolite group, varieties

Hardness: 3.5–5.5 **Streak:** White

Primary Occurrence

Environment: Shores, rivers, outcrops

What to Look for: Small, light-colored, brittle and often fibrous or radial crystals filling cavities in dark volcanic rocks

Size: Individual crystals tend to be small, under an inch; crystal groups or masses may be up to fist-size or rarely larger

Color: Varies; typically colorless to white or gray, tan to brown, yellowish to orange, pink to salmon, green; can be banded

Occurrence: Common to rare, depending on species

Notes: Several zeolites in the region—including natrolite, mesolite, and thomsonite—share the commonly occurring needle-shaped crystal form that zeolites are best known for. Except in the most finely formed examples, these minerals can appear so similar that distinguishing them won't be possible outside of a lab, and in fact many of these minerals occur intimately intergrown together in the same crystal cluster. In these cases, when a radial cluster of brittle, elongated crystals is the only readily identifiable trait, you'll have to settle for simply labeling a specimen "zeolite." While some zeolites can be found as loose water-worn pebbles weathered free from their host rock, most will be found still tightly embedded. When they haven't completely filled the cavity, other minerals like chlorite or calcite can fill the void, often encasing the zeolite within them. While individual species can be difficult to identify, spotting zeolites in general is not hard; few other soft, cavity-filling minerals share the fibrous, radiating shapes of most Lake Superior zeolites.

Where to Look: Most rivers along Minnesota's shoreline while yield samples, especially laumontite. Basalt and gabbro exposures around Amnicon Falls in northern Wisconsin are known for specimens, as are the shores along the Keweenaw Peninsula from Eagle Harbor to its tip.

Cluster of analcime crystals

Ball-like crystal shape

Large crystal

Iron-stained crystals

Basalt host rock

Crust of poorly formed analcime crystals

Crude river-worn analcime masses (tan)

Zeolite group, Analcime

Hardness: 5–5.5 **Streak:** White

Primary Occurrence

Environment: Shores, rivers, outcrops

What to Look for: Small, colorless or brownish angular, ball-shaped crystals clustered within cavities in basalt

Size: Individual crystals are small, rarely larger than a pea

Color: Colorless to white, tan to brown or reddish brown

Occurrence: Uncommon to rare, depending on location

Notes: Analcime (sometimes called analcite) is one of few zeolites found around Lake Superior that doesn't form as fibrous, needle-like crystals, instead developing as blockier, stout crystals shaped like faceted balls. Crystals are typically found in complex groups, tightly intergrown with each other, and before close inspection may appear superficially similar to calcite or quartz (but calcite is softer while quartz is much harder). Particularly pure specimens are colorless or white and translucent, but iron and other impurities frequently tint specimens to shades of tan to brown or reddish. Crystals are also usually quite lustrous, appearing brightly glassy. As with most zeolites, analcime is typically found embedded within basalt, both within vesicles (gas bubbles) and especially in larger cavities or veins, often alongside calcite. But analcime's crystal shape, combined with its hardness, is distinctive and makes it hard to confuse with anything else unless heavily weathered and worn. Water-worn or massive, poorly formed examples are trickier, but they are often opaque, tan, and may still show glassy faces.

Where to Look: The Copper Harbor, Michigan, area basalts have produced nicely formed reddish crystals. Knife River, Minnesota, is known for clusters of larger crystals in basalt cobbles. Elsewhere, beach finds can rarely turn up, usually appearing as angular tan masses identifiable primarily by hardness.

Basalt with laumontite-filled vesicles (tan-orange)

Highly weathered, crumbly basalt

Laumontite in quartz geode

Laumontite in prehnite

Calcite

Broad, elongated crystals

Large, crumbly laumontite mass

Primary Occurrence

Zeolite group, Laumontite

Hardness: 3.5–4 **Streak:** White

Environment: Shores, rivers, pits

What to Look for: White to salmon-colored, opaque, elongated crystals in basalt, often coarsely formed and very brittle

Size: Individual crystals may be up to an inch or two; masses can be fist-size and rarely larger

Color: White to gray; often pink to salmon, yellowish brown

Occurrence: Common to very common, depending on location

Notes: Among Lake Superior's most common zeolites is laumontite, a commonly salmon-pink mineral that formed in large amounts within the region's basalt formations, often with or on calcite and prehnite. Like many other zeolites, laumontite primarily develops as elongated crystals arranged into fan-shaped clusters. But laumontite's crystals tend to be coarser and blockier, often grown large enough that their square- or diamond-shaped cross-section can be observed, which will be helpful in identification. In addition, laumontite's crystal clusters are often tightly packed and more poorly organized than in other zeolites. All zeolites contain some amount of water trapped within their crystal structure and are subject to dehydration; laumontite is a prime example, and virtually all specimens begin to dehydrate upon exposure to air, becoming brittle and sometimes so crumbly that you can't even collect them. Some highly weathered basalts are so full of laumontite that the whole rock takes on an orange appearance and readily crumbles as well. Wishful thinkers may confuse it with more collectible thomsonite (page 235) but laumontite is softer and far more common.

Where to Look: Laumontite is abundant along Minnesota's shores, especially from Two Harbors to Silver Bay, where virtually any basalt flow will be littered with filled vesicles.

Beach-worn lintonite pebbles

Waxy, translucent surfaces

Thomsonite crystals in lintonite

Lintonite nodules within water-worn host basalt

Zeolite group, "Lintonite"

Hardness: 5–5.5 **Streak:** White

Primary Occurrence

Environment: Shores, rivers

What to Look for: Small, rounded, gray-green translucent pebbles, sometimes embedded in basalt

Size: Most specimens are under an inch

Color: Gray-green to bluish green; sometimes pinkish or gray

Occurrence: Rare to very rare, depending on location

Notes: In the 1870s, Laura Linton at the University of Minnesota analyzed some unidentified pebbles found near Grand Marais, Minnesota. Despite being found to have the same chemical composition as thomsonite (page 235), the mineral's structural and optical properties were different enough that it was classified as a new, distinct mineral, named "lintonite" in her honor. Today, we know that lintonite actually is thomsonite, but a particularly odd variety that formed massively and without a distinct crystal structure, rather than as the typical fine, needle-like crystals normally seen in the mineral. Usually a gray-green coloration and generally translucent with waxy or softly lustrous surfaces, lintonite can be an attractive collectible, usually found as loose pebbles weathered free of their host basalt but occasionally as nodules (round masses) still embedded in their vesicles (gas bubbles). Rarely, they can contain signs of typical pinkish thomsonite crystals within them, sometimes revealed at their center when cut open. They are easily confused with water-worn chalcedony (page 103) and prehnite (page 187), especially the variety of prehnite known as "Patricianite," but both of those minerals are quite a bit harder.

Where to Look: The beaches and rivers near Grand Marais, Minnesota, are the primary place to look. They could also rarely turn up on the Keweenaw Peninsula beaches.

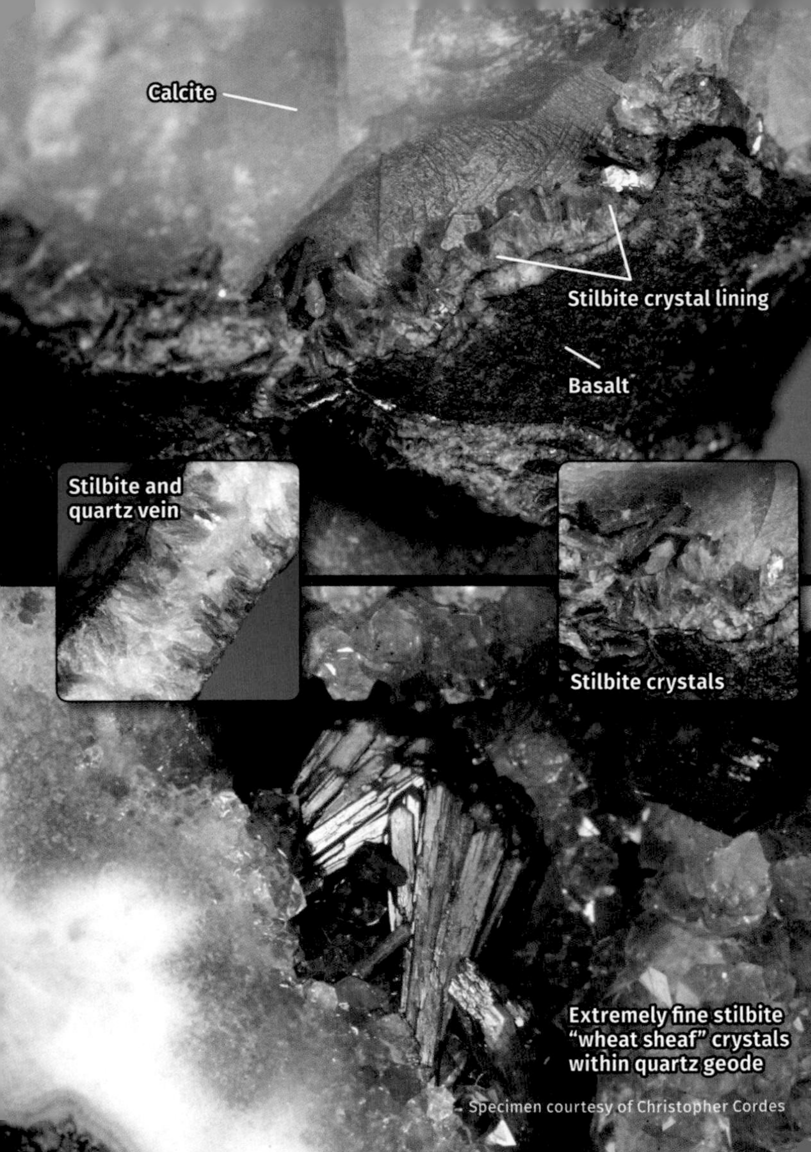
Calcite
Stilbite crystal lining
Basalt
Stilbite and quartz vein
Stilbite crystals
Extremely fine stilbite "wheat sheaf" crystals within quartz geode
Specimen courtesy of Christopher Cordes

Zeolite group, Stilbite

Hardness: 3.5–4 **Streak:** White

Primary Occurrence

Environment: Shores, rivers, outcrops

What to Look for: Thin, plate-like crystals with wedge-shaped tips, often arranged into radial clusters or crusts in basalt

Size: Individual crystals are small, usually around ⅛ inch; clusters or veins may measure up to several inches

Color: Usually pink to orange or reddish brown; sometimes gray

Occurrence: Uncommon to rare, depending on location

Notes: Stilbite is a zeolite mineral that is probably more common in the Lake Superior region than it seems. Fine crystals are attractive and highly collectible but are found much more sparingly than those of other zeolites. Instead, stilbite is certainly more abundant as nondescript orange coatings or veins, which are usually misidentified as the much more common laumontite (page 229). But while rare, stilbite's small crystals are distinctively shaped, appearing as flat, elongated plates with wedge-shaped tips, sometimes arranged into narrow fan-shaped clusters often described as a wheat-sheaf shape. Crystal groups can also line the walls of a cavity or crack, usually in basalt, and have a flaky structure when broken. In all cases, stilbite typically shows a softly reflective pearly luster and is most often orange in color, making for a stark contrast with the white calcite or quartz with which it commonly occurs. Laumontite is the mineral you'll likely confuse for stilbite due to their similar color, but laumontite forms coarser, squarer crystals that are usually dull and opaque.

Where to Look: Minnesota's lakeside basalt formations will be one of the primary areas to find stilbite; adjacent rivers are often the most lucrative, from Knife River to Grand Marais. The northernmost portion of the Keweenaw Peninsula, Michigan, is also known for small crystals.

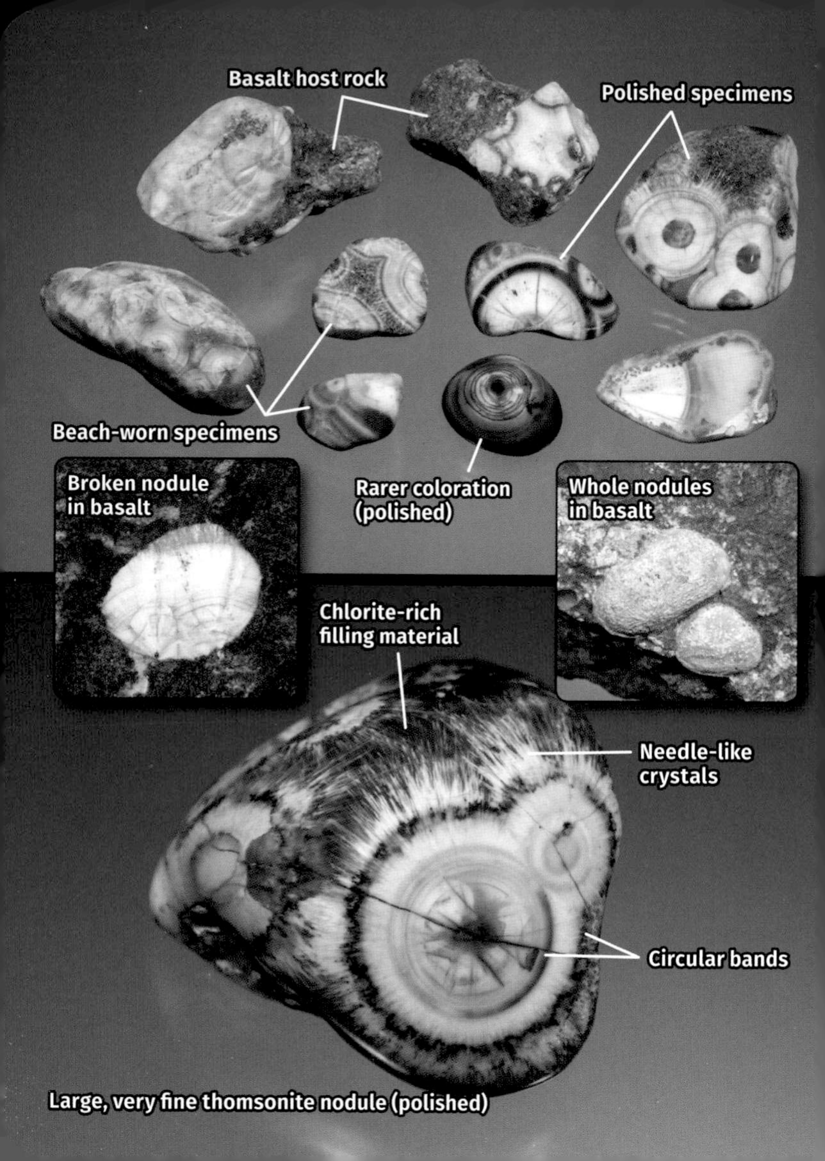

Basalt host rock

Polished specimens

Beach-worn specimens

Rarer coloration (polished)

Broken nodule in basalt

Whole nodules in basalt

Chlorite-rich filling material

Needle-like crystals

Circular bands

Large, very fine thomsonite nodule (polished)

Zeolite group, Thomsonite

Hardness: 5–5.5 **Streak:** White

Primary Occurrence

Environment: Shores, rivers, outcrops

What to Look for: Nodules of white or pink needle-like crystals arranged into circular "sprays," usually in basalt

Size: Thomsonite nodules are rarely as large as a golf ball

Color: White to pink, often with green to bluish green bands

Occurrence: Uncommon to rare, depending on location

Notes: One of Lake Superior's best known and most popular collectibles is thomsonite, a zeolite famous for its attractive circular patterns. Lake Superior is one of the only places in the world where the usually delicate, needle-like crystals of thomsonite form in dense, tightly intergrown nodules (rounded masses) suitable for use in jewelry. Its crystals form in radial clusters, sometimes fan-shaped but more often growing outward in all directions, making for circular "eyes." The eye-like structures can take a variety of colors and appearances, but a common feature are its rings of color caused by influxes of impurities during crystal growth, such as red iron-bearing minerals and green chlorite (page 111). Whole nodules, as well as ones that have been broken open, are found embedded in cavities in basalt and diabase, and even though it is fairly soft, thomsonite nodules are robust enough to survive as beach-worn pebbles, too. But thomsonite specimens are rarely pure; when analyzed, they almost always contain some needles of similar zeolites, chiefly mesolite and natrolite. Don't confuse it with laumontite (page 229), which is softer and more common.

Where to Look: The famous Thomsonite Beach and Terrace Point areas near Grand Marais, Minnesota, are highly trafficked but still lucrative. The tip of the Keweenaw Peninsula, especially the Copper Harbor area, is also good.

GLOSSARY

Aggregate: An accumulation or mass of crystals

Alkaline: Substances containing alkali elements such as calcium, sodium, and potassium; having the opposite properties of acids

Alter: Chemical changes within a rock or mineral due to the addition of mineral solutions

Amphibole: A large group of important rock-forming minerals commonly found in granite and similar rocks

Amygdule: A nodular mineral formation within a vesicle; often said to be almond shaped

Associated: Minerals that often occur together due to similar chemical traits or similar processes of formation

Band: An easily identified layer within a mineral

Bed: A large, flat mass of rock, generally sedimentary

Botryoidal: Crusts of a mineral that formed in rounded masses, resembling a bunch of grapes

Breccia: A coarse-grained rock composed of broken angular rock fragments solidified together

Chalcedony: A massive, microcrystalline variety of quartz

Cleavage: The property of a mineral to break along the planes of its structure, which reflects its internal atomic organization; referred to in terms of shape or angles

Compact: Dense, tightly formed rocks or minerals

Concentric: Circular, ringed bands that share the same center, with larger rings encompassing smaller rings

Conchoidal: A circular shape, often resembling a half-moon; generally referring to fracture shape

Crust: The rigid outermost layer of the earth

Crystal: A solid body with a repeating atomic structure formed when an element or chemical compound solidifies

Cubic: A box-like structure with sides of an equal size

Dehydrate: To lose water contained within

Druse: A coating of small crystals on the surface of another rock or mineral

Dull: A mineral that is poorly reflective

Earthy: Resembling soil; dull luster and rough texture

Effervesce: When a mineral placed in an acid gives off bubbles caused by the mineral dissolving

Eruption: The ejection of volcanic materials (lava, ash, etc.) onto the earth's surface

Face: A distinct, typically smooth surface of a crystal derived from a mineral's structure

Feldspar: An extremely common and diverse group of light-colored minerals that are most prevalent within rocks and make up the majority of the earth's crust

Fibrous: Fine, rod-like crystals that resemble cloth fibers

Fluorescence: The property of a mineral to give off visible light when exposed to ultraviolet light radiation

Fracture: The way a mineral breaks or cracks when struck, often referred to in terms of shape or angles

Fault: A broad, planar crack or break in a rock, usually caused by shifting; faults can interrupt or disrupt layers and cause shifting of banded patterns

Geode: A hollow rock or mineral formation, typically exhibiting a very round, ball-like external shape and interior walls lined with minerals, namely quartz and calcite

Glacier: A large body of ice that moves under its own weight in conjunction with freezing and melting; glaciers are a defining characteristic of Ice Ages

Glassy: A mineral with a reflectivity similar to window glass, also known as "vitreous luster"

Gneiss: A rock that has been metamorphosed so that some of its minerals are aligned in parallel bands

Granitoid: Pertaining to granite or granite-like rocks

Granular: A texture or appearance of rocks or minerals that consist of grains or particles

Hexagonal: A six-sided structure

Host: A rock or mineral on or in which other rocks and minerals occur

Hydrous: Containing water

Igneous rock: Rock resulting from the cooling and solidification of molten rock material, such as magma or lava

Impurity: A foreign mineral within a host mineral that often changes properties of the host, particularly color

Inclusion: A mineral that is encased or impressed into a host mineral

Iridescence: When a mineral exhibits a rainbow-like play of color, often only at certain angles

Lava: Molten rock that has reached the earth's surface

Lava flow: An igneous rock formation that retains the general shape or appearance of when it initially hardened from flowing molten lava

Ledge rock: Hard, exposed bedrock, typically seen on Lake Superior's shores as sheets of igneous rock

Luster: The way in which a mineral reflects light off of its surface, described by its intensity

Magma: Molten rock that remains deep within the earth

Malleable: Easily bent without breaking, as in pure metals

Massive: Mineral specimens found not as individual crystals but rather as solid, compact concentrations; in geology, "massive" is rarely used in reference to size

Matrix: The rock in which a mineral forms; see host

Metamorphic rock: Rock derived from the alteration of existing igneous or sedimentary rock through the forces of heat and pressure

Metamorphosed: A rock or mineral that has already undergone metamorphosis

Mica: A large group of minerals that occur as thin flakes arranged into layered aggregates resembling a book

Microcrystalline: Crystal structure too small to see with the naked eye

Mine dump: Piles of waste rock left behind at mine sites. This rock was produced when mine shafts were dug to access valuable ores and may or may not contain collectible minerals.

Mineral: A naturally occurring inorganic chemical compound or native element that solidifies with a definite internal crystal structure

Native element: An element found naturally uncombined with any other elements (e.g. copper)

Nodule: A rounded mass consisting of a mineral, generally formed within a vesicle or other cavity; a mineral specimen formed in this way is said to be nodular

Octahedral: A structure with eight faces, resembling two pyramids placed base-to-base

Opaque: Material that lets no light through

Ore: Rocks or minerals from which metals can be extracted

Oxidation: The process of a metal or mineral losing electrons to another element or material, often when combining with oxygen, which can produce new colors or minerals

Pearly: A mineral with reflectivity resembling that of a pearl

Placer: Deposit of sand containing dense, heavy mineral grains at the bottom of a river or a lake

Pyroxene: A group of hard, dark, rock-building minerals that make up many dark-colored rocks like basalt or gabbro

Radiating: Crystal aggregates growing outward from a central point, often resembling the shape of a paper fan

Rhombohedron: A six-sided shape resembling a leaning or skewed cube

Rock: A massive aggregate of mineral grains

Rock-builder: Refers to a mineral important in rock formation, usually in reference it igneous rocks

Schiller: A mineral that exhibits internal reflections or "flashes" from within its structure when rotated in bright light, often showing an interplay of white, yellow, or blue

Schist: A rock (usually sedimentary) that has been metamorphosed so that most of its minerals have been concentrated and arranged into parallel layers

Sediment: Fine particles of rocks, minerals, or organic matter deposited by water or wind (e.g. sand)

Sedimentary rock: Rock derived from sediment being cemented together

Silica: Silicon dioxide; pure silica crystallizes to form quartz; silica contributes to the makeup of thousands of minerals

Slag: Waste material produced in the processing of ores; often with a rocky, glassy, or "bubbly" texture and sometimes with a nearly metallic luster

Species: A mineral distinguished from others by its unique chemical and physical properties.

Specific gravity: The ratio of the density of a given solid or liquid to the density of water when the same amount of each is used (e.g. the specific gravity of copper is approximately 8.9, meaning that a sample of copper is about 8.9 times heavier than the same amount of water)

Specimen: A sample of a rock or mineral

Stalactite: A cone-shaped mineral deposit grown downward from the roof of a cavity; sometimes described as icicle-shaped. Formations in this shape are said to be stalactitic

Striated: Parallel grooves in the surface of a mineral

Tabular: A crystal structure in which one dimension is notably shorter than the others, resulting in flat, plate-like shapes

Tarnish: A thin coating on the surface of a metal, often differently colored than the metal itself (see oxidation)

Tectonic plate: The enormous sheets of rock that make up the earth's crust and upon which the continents and oceans reside. Tectonic plates move slowly as a result of the extreme heat below them, causing earthquakes and volcanoes, and forming mountains and oceans

Translucent: A material that lets some light through

Transparent: A material that lets enough light through so one can see what lies on the other side

Vein: A mineral that has filled a crack or similar narrow opening in a host rock or mineral

Vesicle: A cavity created in an igneous rock by a gas bubble trapped when the rock solidified; a rock containing vesicles is said to be vesicular

Volcano: An opening, or vent, in the earth's surface that allows volcanic material such as lava and ash to erupt

Vug: A small cavity within a rock or mineral that can become lined with different mineral crystals

Waxy: A mineral with a reflectivity resembling that of wax

Weathering: The process of rocks and minerals being worn down by exposure to the elements, such as wind, rain, waves, ice, glaciers, and naturally occurring chemicals

Zeolite: A group of similar minerals with very complex chemical structures that include elements such as sodium, calcium, and aluminum combined with silica and water and that typically form within cavities in basalt as it is affected by mineral-bearing alkaline groundwater

LAKE SUPERIOR ROCK SHOPS AND MUSEUMS

Minnesota

AGATE CITY ROCKS AND GIFTS
721 7th Ave
Two Harbors, MN 55616
(218) 834-2304

BEAVER BAY ROCK SHOP
1003 Main St
Beaver Bay, MN 55601
(218) 226-4847

GRAND MARAIS ROCK SHOP
1821 W Hwy 61
Grand Marais, MN 55604
(651) 485-9973

Michigan

A. E. SEAMAN MINERAL MUSEUM *(a must-see museum)*
1404 E Sharon Ave
Houghton, MI 49931
(906) 487-2572
museum.mtu.edu

CLIFFS SHAFT MINE MUSEUM
501 W Euclid St
Ishpeming, MI 49849
(906) 487-2572

KEWEENAW GEM AND GIFT
912 Razorback Dr
Houghton, MI 49931
(906) 482-8447

Michigan *(continued)*

PICTURED ROCKS INTERPRETIVE CENTER
100 W Munising Ave
Munising, MI 49862
(906) 387-1900

PROSPECTOR'S PARADISE
3 miles north of Calumet
US Hwy 41,
Allouez, MI 49805
(906) 337-6889

QUINCY MINE HOIST *(rock shop and mine tours)*
49750 US Hwy 41
Hancock, MI 49930
(906) 482-3101

RED METAL MINERALS
202 Ontonagon St
Ontonagon, MI 49953
(906) 884-6618

ROCK KNOCKER'S ROCK SHOP
490 North Steel St (US Hwy 41)
Ishpeming, MI 49849
(906) 485-5595

Ontario

AMETHYST GIFT CENTER
400 Victoria Ave E
Thunder Bay, ON P7C 1A5
+1 807 622 6908

AMETHYST MINE PANORAMA *(seasonal, call ahead)*
500 Bass Lake Rd
Shuniah, ON
+1 807 622 6908

Ontario *(continued)*

BLUE POINTS AMETHYST MINE *(seasonal)*
approx. 35 miles east of Thunder Bay on Canada Hwy 11/17
5 Rd N, Shuniah,
ON P0T 2M0

DIAMOND WILLOW AMETHYST MINE *(seasonal, call ahead)*
approx. 35 miles east of Thunder Bay on Canada Hwy 11/17
5 Rd N, Shuniah,
ON P0T 2M0
+1 807 627 5515

PURPLE HAZE AMETHYST *(rock shop)*
22 Knight St
Thunder Bay, ON P7A 5M2
+1 807 345 6444

BIBLIOGRAPHY AND RECOMMENDED READING

Books About Lake Superior Minerals

Dott, Robert H. and Attig, John W. *Roadside Geology of Wisconsin*. Missoula: Mountain Press Publishing, 2004.

Heinrich, E. W. and Robinson, George W. *Mineralogy of Michigan*. Houghton: Michigan Technological University, 2004.

Lynch, Dan R. and Lynch, Bob. *Agates of Lake Superior: Stunning Varieties and How They Are Formed*. Cambridge: Adventure Publications, Inc., 2011.

Ojakangas, Richard W. *Roadside Geology of Minnesota*. Missoula: Mountain Press Publishing, 2009.

Ojakangas, Richard W. and Matsch, Charles L. *Minnesota's Geology*. Minneapolis: University of Minnesota Press, 1982.

Pye, E. G. *Roadside Geology of Ontario: North Shore of Lake Superior*. Sudbury: Ontario Ministry of Northern Development and Mines, 1997.

Robinson, George W. and LaBerge, Gene L. *Minerals of the Lake Superior Iron Ranges*. Houghton: Michigan Technological University, 2013.

General Reading

Bonewitz, Ronald Louis. *Smithsonian Rock and Gem*. New York: DK Publishing, 2005.

Chesteman, Charles W. *The Audubon Society Field Guide to North American Rocks and Minerals*. New York: Knopf, 1979.

Johnsen, Ole. *Minerals of the World*. New Jersey: Princeton University Press, 2004.

Mottana, Annibale, et al. *Simon and Schuster's Guide to Rocks and Minerals*. New York: Simon and Schuster, 1978.

Moxon, Terry. *Studies on Agate: Microscopy, Spectroscopy, Growth, High Temperature and Possible Origin*. Doncaster: Terra Publications, 2009.

Pellant, Chris. *Rocks and Minerals*. New York: Dorling Kindersley Publishing, 2002.

Pough, Frederick H. *Rocks and Minerals*. Boston: Houghton Mifflin, 1988.

Robinson, George W. *Minerals*. New York: Simon & Schuster, 1994.

INDEX

ABOUT THE AUTHORS

Dan R. Lynch has a degree in graphic design with an emphasis in photography from the University of Minnesota Duluth. But before his love of art and writing came a passion for nature. His interest in rocks and minerals was cultivated growing up in his parents' rock shop near Lake Superior's shore, learning rock and mineral identification first-hand. Initially working with his father, Bob Lynch, Dan has since written more than 20 books about rocks and enjoys educating his readers in a straightforward, easy-to-understand manner. Dan's photography complements his books and he takes special care to ensure that his photographs always honestly represent the specimens pictured. By introducing a complex subject like geology in a clear, concise way, and by starting at the beginning, he hopes to excite readers and spark a lifelong interest in the amazing science underfoot. He lives in Madison, Wisconsin, with his wife, Julie, where he works as a classical numismatist and a writer.

ABOUT THE AUTHORS *(continued)*

Bob Lynch is a jeweler and lapidary living and working in Two Harbors, Minnesota. His experience in cutting and polishing rocks and minerals began in 1973, when he desired more variation and control over which gemstones he could use in his jewelry. After years of working with turquoise and malachite, he moved from Douglas, Arizona, to Two Harbors in 1982, and his eyes were opened to Lake Superior's entirely new world of minerals—especially agates. In 1992, Bob and his wife, Nancy, whom he taught the art of jewelry making, acquired Agate City Rock Shop, a family business founded by Nancy's grandfather, Art Rafn, in 1962. Since the shop's revitalization, Bob has made a name for himself as a highly acclaimed agate polisher and as an expert resource for curious collectors seeking advice. Today, the semi-retired jewelers still own and operate Agate City, open throughout the summer months and most weekends year-round. When he's not creating jewelry, Bob enjoys trap shooting and visiting his sons.

NOTES

NOTES